The Physical Geography (Geomorphology)

of

WILLIAM MORRIS DAVIS

THE PHYSICAL GEOGRAPHY
(Geomorphology)

of

WILLIAM MORRIS DAVIS

Compiled, Illustrated, Edited and Annotated

by

Philip B. King

and

Stanley A. Schumm

GEO BOOKS

1980

The Physical Geography (Geomorphology) of William Morris
Davis   is compiled, illustrated, edited and annotated by
Philip B. King & Stanley A. Schumm.

Copyright Philip B. King & Stanley A. Schumm 1980.

First published by Geo Abstracts Limited
in their GEO BOOKS series,   May 1980

ISBN Hardback (with appendix) 0 86094 046 2
     Paperback (no appendix)   0 86094 047 0

Geo Abstracts, Norwich, England

CONTENTS

Contents

LIST OF FIGURES

List of figures

List of figures

Acknowledgments

The following people have provided encouragement and useful suggestions for preparation of the annotations:
  Dr. Malcolm Anderson, University of Bristol;
  Prof. Arthur L. Bloom, Cornell University;
  Prof. T.C. Chamberlain, Colorado State University, and
  Prof. Richard J. Chorley, University of Cambridge.
  Professor Bloom reviewed the entire text, and provided valuable advice based on his thorough grasp of geomorphic concepts. Several of his comments were incorporated into the notes.

Senior Editor's Preface

In 1912, William Morris Davis, who had long been a
professor at Harvard University, became professor emeritus,
and for a time devoted himself to various research projects, such as a study of coral reefs in the South Pacific.
From the early twenties and for the remainder of his life,
until 1934, he was lecturer and visiting professor at many
western schools. Although his great system of geomorphology (or 'physical geography' as he always called it)
had been evolved during his Harvard years, in many ways
the period after his retirement was the most influential
of his career.

Whereas at Harvard he was, to his students, a crotchety
and unhelpful taskmaster (so G.R. Mansfield told me),
during his career in the western schools he lectured with
fewer inhibitions regarding his philosophy; these lectures,
and his many personal contacts, profoundly influenced a
whole generation of younger earth scientists. (Although
he always considered himself to be a 'geographer' his
greatest impact was on young geologists.) All these
younger men knew instinctively that Davis was a great man
and teacher, at whose feet they should sit and worship.

I knew this, for one, because at the same time I was
becoming acquainted with other famous geologists - whose
fame so often diminished as one approached, and who often
seemed to operate with the most trivial and petty of
motives. Davis was never trivial or petty! All his faults
and virtues had epic proportions! But this aside, Davis's
inspiration to me went beyond his own specialty, for he
introduced me to deductive science. Already, in the
nineteen-twenties, geologists were off in pursuit of a more
exact basis for their science, couched in terms of
mathematics, physics, and chemistry - a trend which
oppresses us even more today, and for which I knew from
the beginning that I had no talent whatsoever. But Davis
showed how many secrets could be unlocked by simple
rationalization and deduction, and offered the hope that,
for a few decades at least, there was still an opportunity
for a constructive career in non-mathematical geology.

My first contact with Davis was when I was doing a
semester of graduate study at the University of Iowa in
the autumn of 1925, when he stopped off briefly to lecture
on the Basin Ranges and on Coral Reefs.

In the following year, by good fortune, I was able to
renew the acquaintance at the University of Texas, when
Davis was visiting professor during the winter term of
1926-1927, and I was a young instructor. I did not
register for his course, but I attended his lectures
faithfully, taking the notes which have been written up in

this volume. I answered some of his examinations, and I still have in my possession the papers, graded and vigorously corrected by him.

For the most part, Davis's influence at Texas fell on sterile ground; students at Texas who took his course were simply not prepared for him. One day, after a long exposition of a region with cuesta topography, he asked, "Where is it?", and one student said, "It sounds like the country down around Houston". Actually, the region was southeastern England, whose resemblance to anything in the Texas Coastal Plain was remote. His greatest influence at Texas, then, was on post-graduate geologists - myself for one, and W.S. Adkins for another, both of us at the time preparing geological reports on West Texas. Davis opened our eyes to the principles of arid erosion, not widely understood at the time, which bore directly on our work - although we were continuously frustrated by his insistence on a form of concise 'geographical description', supposedly the ultimate objective of his whole system, but actually an absurd and unworkable formula, little used by Davis himself in his own writings.

My later contact with Davis was at the University of Arizona, where he was visiting professor during the spring semester of 1930, and I was again an instructor. I did not attend his classes this time, and our relations were entirely personal and social, but knowing him again helped me much to fill in the chinks of my conception of his philosophy.

He and Mrs. Davis arranged many picnics for the geology faculty to places of geomorphic interest in the desert around Tucson, at which Davis played the great man and showman. At other times I had opportunities to discuss with him some of his work in progress - on desert erosion, basin-range topography, and processes of cave formation. Once I was invited on a field trip when Douglas Johnson was a visitor, to see the desert terrain around Tucson. I recall that Johnson told Davis he thought Davis's terms, the 'King formations', 'King folds', 'Powell surface', 'Gilbert fault blocks', and 'Louderbacks', were "rather trivial" - a liberty that only another 'great' could take with Davis. I also took Davis on some field trips of my own, where there was an odd contrast between the views of a theoretical geomorphologist and that of a practical field geologist. It was on one on these trips that (Quaker puritan that he was) he denounced the highway signs south of Tucson, advertising night spots in Nogales, as "an invitation to a debauch".

From 1930 until 1962, my Davis notes were carefully preserved, with the hope that someday something could be done with them. A few topics had been written up in better form than the rest, but most of them remained in the rough form in which they had been put together in Austin and Tucson, years before. In 1962, a break in my work for the U.S. Geological Survey offered an opportunity to work up the material in better form. This work was a labor of love, intended mainly for my personal reference, but I did make copies of the typed pages, that I deposited in the three libraries of the Geological Survey.

As time went on, I discovered that these library copies had been much used, and I was urged by many geomorphologists to make them available in a more permanent publication - requests that have continued until now. It has only been recently, after my retirement from the Geological Survey, that the time was available to do the necessary typing and editing to make this possible, and the results are presented herewith, with the able help and counsel of Stanley Schumm.

During the nearly half-century gap between the original note-taking and the present, Davis's concepts have gone into gradual eclipse among geomorphologists, and many of them, it is true, now seem quaint, old-fashioned, and over-simplified. But many of the great principles on which they were based, though often overlooked today, are still valid - especially the concept of a continual evolution of land-forms from the past, through the present, into the future. Valid or not, the record is at least worth preserving as a statement of the methods used by Davis, directly from the master himself.

The text presented herewith is based largely on my original notes on the Davis lectures, with only minor editing and rewriting. In addition, I acquired from Davis in Tucson in 1930 a 42-page text with 3 pages of errata, labeled 'University of California, Summer Session 1929, Geology 102, Outline No. 1', covering a course given at Berkeley the year before. This covers much the same ground as my own notes and amplifies them in places; where appropriate, these parts are worked into my text. In general, I have tried to retain as far as possible the statements made by Davis himself. In a few places Davis presented remarks on the historical background of geo-morphological studies, as on the Great Plains of Montana, the Basin Ranges, and arid landforms; I have amplified these somewhat from the available record. Also, for the section on Coral Reefs, my original notes were inadequate, so I have expanded them from Davis's own publications. Other than these, my own comments are given in notes at the end, along with many other comments contributed by Stanley Schumm.

Actually, my own personal interest is not in the theories of geomorphology, but in landscape drawing and diagramming, a subject in which Davis was also a master; a secondary major objective of this book is to present a large selection of the Davis drawings and diagrams. Many of the smaller figures given here are my copies of Davis's blackboard sketches made during the lectures. Besides these, I have endeavored to collect and reproduce as many as possible of the superb landscape drawings and diagrams that accompanied Davis's publications. These figures, I found, were in widely scattered publications, and had never before been collected in one place. I have reproduced here all those which I judge to be significant; those which I have not used I believe to be too trivial to be worth considering. Many of these figures illustrate specific points in the text, but others which do not bear directly on the text are included for their own sake. It is hoped that this collection will inspire other geologists and geomorphologists to continue to practice the art.

Philip B. King
Menlo Park CA
1979

Photo 1   W.M.Davis about 1912,   the year of the publication
of *Die Erklärende Beschreibung der Landformen.*

Junior Editor's Preface

Sometime late in 1966 I discovered in the Geological Survey
library in Denver an intriguing document. It consisted of
a set of illustrated lecture notes from a course entitled
Physical Geography, which was presented by W.M. Davis in
1927 at the University of Texas and an outline of a similar
course presented in 1929 at the University of California
at Berkeley. The notes were of considerable interest to me
at that time because they showed, among other things, that
Davis was discussing and had accepted parallel-slope
retreat and pediments during these lectures. My knowledge
of his work was primarily second-hand and through his early
publications (1890-1906), as collected by D.W. Johnson in
the volume Geographical Essays, and everything that I had
read indicated that Davis was the advocate of declining
slope retreat. Nevertheless, the notes showed that he had
accepted the concept of parallel slope retreat before 1927.

The original Davis papers are very long and sometimes
tedious, but the notes provided an opportunity to review
his geomorphology in brief. The only other such review was
published in German by Davis in 1912 as *Die Erklärende
Beschreibung der Landformen*. Subsequently, I corresponded
with Philip King in order to acquire a personal copy of
the notes. Through the years I remembered my excitement in
finding this compilation, and I regretted that the notes
were not available to a wider readership. In 1976 King
and I agreed to do something about the publication of the
notes, and the results are before you.

The notes have been edited both by King and myself and
the material in a 42 page outline which was prepared by
Davis for the University of California course was partly
incorporated into King's notes. The outline was used to
fill out those parts of King's notes that dealt with topics
that were given a very brief treatment in 1927.

The notes contain some material that can now be considered trivial and some that is incorrect, and for this
reason the text is annotated. Nevertheless, the opportunity
to present in one publication the mature considerations of
an eminent geomorphologist led us both to the conclusion
that the notes should be made available to the geomorphic
public. Much of the value of the notes are the illustrations prepared by King from Davis's blackboard sketches and
from his published works. Few geomorphologists today have
this facility with pen and ink.

Perhaps the main reason that we believe that the notes
should be made generally available is, as Chorley stated
(1965),

"that what is most easily available to students today
as the 'essential' teachings of Davis are certain of

his essays written prior to 1909 and the writings of his most influential students."

Nevertheless, Davis

"showed remarkable versatility after the age of seventy, modifying his views on peneplanation and the youthful stage, recognizing the lack of real differences between many humid and arid landforms, and acknowledging - the difficulty of applying simple cyclical notions to an area of active orogeny."

In fact, he was intuitively aware of the significance of the hydraulic concept of roughness and continuity (Chow, 1959).

Much of the young Davis appears in these notes, but his recognition of the complexity of the landscape and the need for alternative explanations is also present in his lectures delivered in his 77th year. We hope this abbreviated version of his system of geomorphology will be of interest and may partly rehabilitate his reputation.

The concept of the erosion cycle as advanced by Davis has been attacked and defended by many earth scientists, but nevertheless, Davis produced a revolution in geomorphology by providing a paradigm from which to evaluate the landscape of this planet. Certainly Davis cannot be blamed if his followers attempted to force nature to conform to this paradigm.

An appropriate inclusion in this preface is a quote from Davis's outline prepared for the University of California course. In it he tells us what the course is about and how to see the landscape.

"Land Physiography, with which this course is concerned, treats chiefly the present forms of the continents and islands, their origins and changes; it thus treats the inorganic side of Geography. The description of a landscape may omit all theoretical explanation and state only the actual facts of occurrence that may be directly observed. Such a description may be called empirical, as it is based on observational experience. Or the attempt may be to describe the visible facts of a landscape as far as possible in terms of their origin. Such description may be called rational, genetic, or explanatory. Explanatory description will be emphasized in this course.

"For example, in the case of what appears to be an up-arched highland, more or less dissected by its rivers, but not as yet sufficiently dissected to obliterate completely all parts of its highland surface, an explanatory description involves the explicit statement of three items: 1) the preexistent form, before the up-arching took place, 2) the effect of the up-arching, and 3) the work of erosional agencies upon the up-arched mass during and after its up-arching.

"The first of these three items itself calls for a three-item statement: A) the structure of the underlying rock mass, B) the nature of the agencies that have worked on it, and C) the stage reached in their work when the up-arching was initiated. In other words - structure, process, and stage. The ABC of the former cycle make item 1 of the present cycle.

"When the present cycle is so far advanced that all
traces of the former are obliterated, the three-item
description can be shortened by omitting items 1 and
2, and presenting only item 3."
Today the geomorphologist is far less interested in the
explanatory descriptive approach, but nevertheless, the
history of a landscape and of a particular landform must be
understood both in the long and short term (geologic and
engineering time) in order that the effect of man on a
landform can be evaluated and future changes predicted.

Davis would have enjoyed evaluating man's impact, and
he would have been an exceptional writer of Environmental
Impact Statements as well as a convincing expert testifying
in court concerning the results of man's activities on the
landscape.

The lecture notes reflect the situation encountered by
very many professors. The subjects treated in the earlier
part of the course are considered in detail, but as the
end of the semester approaches the time required to
elaborate on the final few topics is too short. Davis too
found time running out, and so the topics of marine and
glacial landforms were given less complete treatment and
the notes terminate abruptly. The first two-thirds of the
notes are of the most interest because of the detail.
Even with King's illustrations the remainder is of much
less value.

Those readers interested in learning more about Davis's
contributions should read Chorley et al (1973), of course,
and should refer in addition to the original publications.
A complete publications list is given by Daly (1945) and
this is reprinted in full as an Appendix to the hard cover
edition of this book. For an explanation of the wide appeal
of Davis's ideas, see Higgins (1975) and Beckinsale (1976).

Stanley A. Schumm,
Fort Collins, Colorado
1979

Photo 2. W.M.Davis photographed three or four years after
giving the lecture course on which this book was
based. The photograph was taken in the field in
California in 1931, two weeks after his 81st.
birthday.

PART 1 THE EROSION CYCLE

The Cycle of Normal Erosion

Development of consequent drainage

In studying the cycle of normal erosion[1]* we will assume a
simple case, to be modified later. A smooth, soil-covered
lowland underlain by nearly uniform rocks is uplifted, upheaved, or upwarped by crust-deforming forces; uplift is
assumed to be instantaneous[2]. A highland is formed sloping
from the main crestline to the sea on either side. Processes which act on this surface are gravity, weathering,
rainfall, wash, and corrasion in stream beds.
   Rain water falling on such a surface will either run
off, evaporate, or sink into the ground. Streams will flow
down the original slope of the land; these are consequent
streams. They will be permanent where the lines of
depression extend below the level of ground water. If the
initial surface includes any basins, they will hold
consequent lakes; if it is broken by scarps, consequent
falls will leap down the scarps (Fig 1).
   Streams that follow the sag lines produced by warping
and upheaval are definite consequent streams. They are fed
by the surface wash and the ground water springs from the
surrounding slopes. Lateral streams which drain into the
sag lines are indefinite consequents[3], or insequents,
because their location cannot be foretold from the shape of
the initial upheaval.
   Indefinite consequents cut steep-sided ravines with
convex profiles between spurs in the valley sides of the
definite consequents of which they are tributary; there
they flow as small torrents. These side ravines often have
hanging mouths, especially if their master consequent is of
large volume. As the lateral ravines are deepened and
broadened the torrents are pushed back to the ravine heads,
and an accordant junction with the master consequent is
developed. The side ravines then become side or lateral

*See notes at end of text.

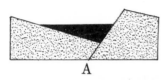 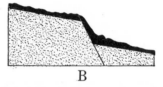

Figure 1. Consequent lake (A), and consequent falls (B)
(from a blackboard sketch).

valleys. Additional gulches are cut back as branches of the
lateral valley near its head; these in time become ravines
and branch valleys. With the increasing development of indefinite lateral valleys, the landscape is diversified.

Valley systems are the most characteristic feature of
the lands. They are linear depressions, excavated by the
streams that drain them, twig joining branch, branch
joining trunk, and all sloping toward the sea. All the
branching streams which unite in a single trunk before
reaching the sea constitute a river system; their valleys
constitute a valley system. Texture of drainage (or
spacing of streams) depends initially on the inequalities
of the uplifted land surface, and later upon the relation
of rainfall, runoff, and sink-in[4]. If the rocks are porous,
like chalk or loose-textured sandstone, much rainfall sinks
into the ground and the runoff is small; streams are then
widely spaced. If the rocks are impervious, runoff is
large, and the streams are closely spaced[5]. Badlands have
extremely close-textured drainage lines[6].

In the early stages of the dissection of an upheaved
landmass, the divides between the streams are vague; on
flat areas the drainage will be indefinite and the runoff
poor[7]. As stream branches deepen and valleys widen, more
and more of the initial surface is transformed into valley-
side slopes, on which discharge is well-directed and rapid.
When the sides of the adjacent widening valleys meet at the
edges, drainage is well-divided. These stream-developed
edges are the divides.

The crestline is sharpened to an edge at the receding
valley heads, but the creeping of soil and rock waste will
round off the edge more or less (Fig 2) - more when the
bedrock is pervious and the rainfall light, less when the
bedrock is impervious and the rainfall heavy[8]. As the
valley heads are slowly worn down to a gentler and gentler
slope, the divides are lowered more and more, and are
usually at the same time more broadly rounded and arched.
When the streams are old, the divides are vague.

The erosion cycle

The time required for the upheaval of a region and its reduction to a plain of degradation is the cycle of erosion.
A cycle of erosion can be divided into a number of stages,
each one characterized by certain advances in erosional
accomplishment. Commonly recognized stages are youth,
maturity, and old age (Fig 3).

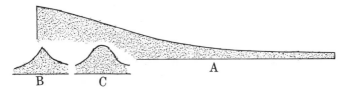

Figure 2. Profile assumed by a stream at grade (A).
B shows the interstream divide which would be produced by stream work alone, while C shows the actual
situation, with the divide modified by weathering and
creep (from a blackboard sketch).

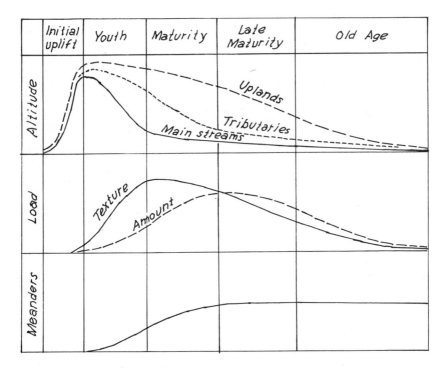

Figure 3. Graphic representation of the erosion cycle
(from a blackboard sketch).

An upheaved highland is <u>young</u> when its rivers, still
torrential in their upper courses, have excavated only
narrow valleys, leaving much of the inter-valley surface
little modified from its initial form by rill-wash and
side-stream dissection (Figs 4, 5A). It is <u>mature</u> when it
is dissected as deeply and abundantly by many branching
valleys as possible, with graded streams and soil-covered
slopes, although the inter-valley summits have as yet lost
little of their initial altitude of upheaval (Fig 5B). Its
general relief is then at maximum value, being strong near

Figures 4 to 7. The erosion cycle in an idealized region. In Fig 4, a region in the old age or peneplain stage of the preceding erosion cycle is actively warped by uplift (Fig 5) after which it remains stable and is progressively eroded. Figs 6A, 6B, and 7A show the progress from early maturity to late maturity. Fig 7B shows the region reduced in relief and approaching the con-

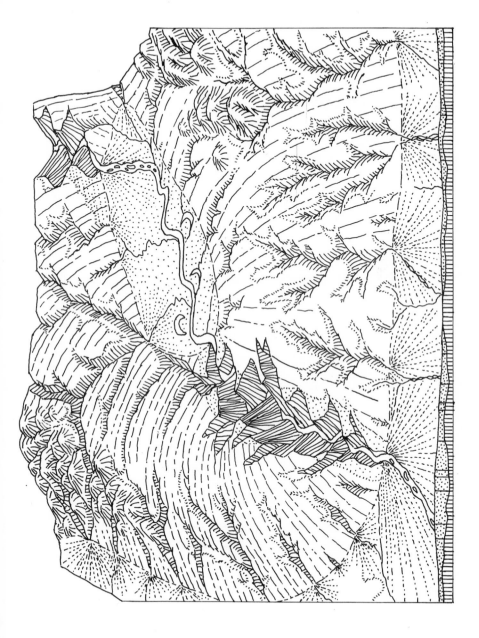

Figure 5

5

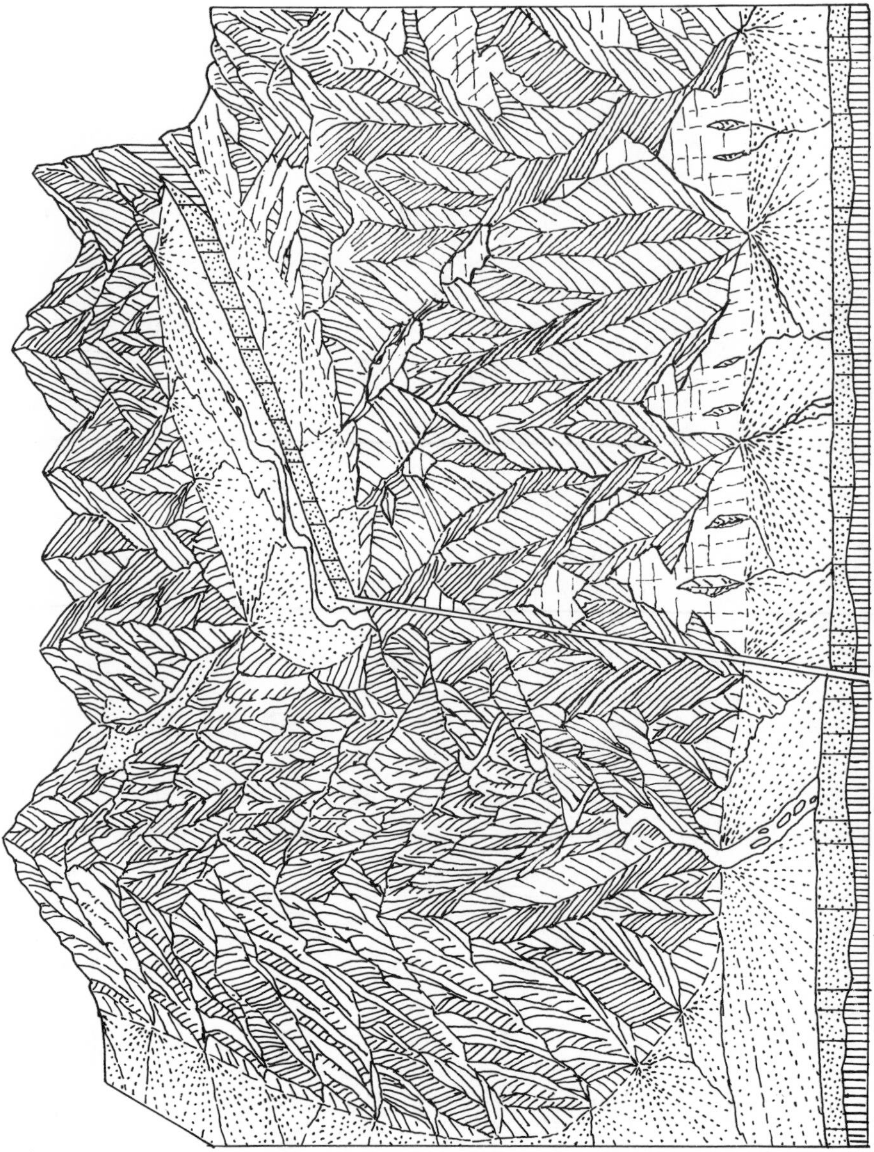

Figure 6A & 6B

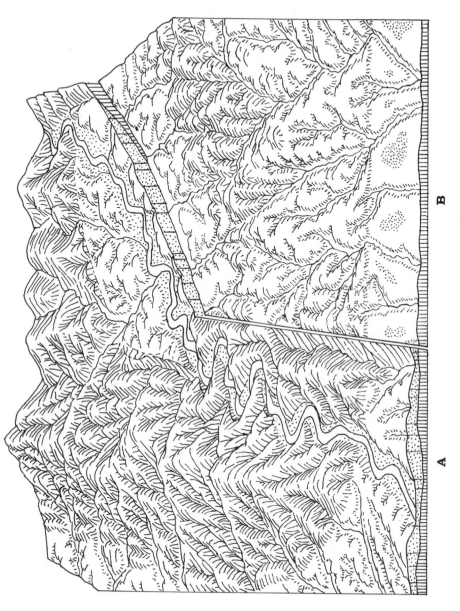

A B

Figure 7A & 7B

the main divide, but low near the seashore. The region
will be late mature when the divides are degraded below
their height at maturity, although still preserving considerable relief (Figs 6A and 6B); the valleys are then
more broadly opened and the stream heads are reduced to
gentler fall than at maturity. The region will be old when
it is everywhere reduced to low relief (Fig 7).

The old age stage is characterized by weakening of
relief and of valley-side slopes; decreasing activity of
all processes of degradation; decreasing rainfall and increasing temperature because of loss in altitude; return to
vaguely defined divides as hills are rounded off to lower
and lower profiles; reduction of all streams to fainter and
fainter fall; refinement of texture of land waste and increase in soil depth.

If a landmass is gradually upheaved (rather than
instantaneously, as in the initial assumption), its rivers
will accomplish a certain measure of valley erosion while
upheaval is still in progress[9]. When upheaval ceases, a
certain length in each lower course will have been graded[10],
while its upper course remains torrential. The more rapid
the upheaval, the harder the bedrock, and the smaller the
rivers, the shorter will be the lower graded course. The
slower the upheaval, the weaker the bedrock, and the larger
the rivers, the longer will be the lower graded course.
This relation may be indicated by saying that, in the
graded course erosion balances upheaval, whereas in the
torrential upper course erosion is unable to balance upheaval.

If an uplifted land stands indefinitely after its upheaval is completed[11]- its valleys will in time be graded
all the way to their heads, and its mountains and hills will
be worn down lower and lower. It will eventually become a
soil-covered plain of degradation, everywhere graded with
respect to baselevel, and drained by slow-flowing rivers of
extremely faint fall in broad-floored valleys that are im-
perceptibly divided from each other by faint swells in the
surface. The level of the ocean, continued under the land,
is the limit of baselevel, below which the land cannot be
eroded by its rivers[12].

At a less advanced stage of degradation, the land will
still possess low, unconsumed hills along the divides and
subdivides between the broad-floored rivers. It will then
be almost-a-plain, or a peneplain[13] (Fig 7). A peneplain
will be hardly above sealevel at its sea margin, but if the
area is large it may attain altitudes of 2000, 3000, or 4000
feet far inland near the river heads, and its residual mounts
and hills may rise still higher, although with gentle slopes.

River morphology

A main river may be divided into three sections: (1) an
upper section where the volume is small, but where it has
much altitude; (2) a middle section of good volume and
medium altitude; and (3) a lower section of large volume
and least altitude[14].

The lower section, with much capacity and little work to
do, soon does that work and grades its shallow valley floor
with respect to normal baselevel at the mouth.

     In the middle section the initial capacity to transport
load by rolling coarse detritus and lifting finer, is
usually greater than the load to be transported. If there
are decreases in the original slope on which the river
flows, so that it there runs too slowly to carry its initial
load, it will lay down part of its load so that it can
carry the remainder. At original increases in slope there
is, on the other hand, a more rapid downcutting. Deposits
of the sort mentioned are likely to be incised when the
river deepens its course farther downstream. The chief
action of corrasion will take place in the middle section.
     By reason of corrading the channel of the river and
thus cutting down a cleft below the initial surface, the
load will be doubly increased - partly by corraded
detritus, and partly by detritus weathered from the cleft
walls. Because of lower gradient the velocity, and thus
the carrying power decrease. Capacity and load, at first
unequal, then approach equality.
     When equality of capacity and load are reached, the
river is graded, in the sense that it can cut no deeper so
long as its load remains unchanged. Nearly all its energy
is then applied to carrying its load[15]. Such a river is
mature, although its valley may still be young. On
maturing, the river tends to be broad and shallow[16].
     The profile assumed by a river at grade is concave up-
ward through most of its course, then convex upward near
its head (Fig 2). The tendency is for the river to develop
a concave profile throughout, with knife-edge divides
between the streams, but these divides are actually rounded
and blunted, due to the dominance of weathering over stream
wash.
     The detritus swept into a stream by wet-weather rills
and rivulets, delivered to its banks by soil creep, and
gathered by corrasion, constitutes its load. Angular
detritus is soon rounded or water-worn by attrition.
Coarse load (gravel and sand) is swept along the bed; finer
load (silt) is lifted by turbulence and carried in suspen-
sion; soluble load is carried in solution. The downstream
carriage of load is called transportation. Transportation
and deposition repeatedly alternate as the load is carried
downstream.
     A river first comes to be at grade near its mouth,
where its large volume gives it much erosive power, and
where its small altitude permits the attainment of grade by
a small amount of erosion. When at grade, the river profile is gently concave, because it needs a slightly steeper
fall as it is followed upstream where the volume is less
and the detrital load is of coarser texture.
     The steep upper course of a young river is actively
engaged in deepening its valley. Owing to high velocity[17],
it sweeps away all but the coarsest detritus, and hence its
current is clear, except where made turbid by muddy
rivulets after a rain. Much bedrock is laid bare in the
channel and on the valley sides. Every slight variation in
the bedrock may affect its corrasion; hence the channel
comes to be of uneven fall and of irregular depth or torrential habit of flow, in which the velocity is less determined by the detrital load than by the internal and

9

external friction[18]. It is drawn out into a slender flow
where the fall is steep, and expands into a broader flow
where the fall is gentle; it will hurry in foaming rapids
where it is steep and slender, and it will loiter in broad
and quiet pools where basins are excavated. In consequence
of faster deepening of the valley near mid-length than at
the head, the profile of the upper course may be somewhat
convex.

The reduction of a torrent to a graded course is
accomplished by the gradual obliteration of some rapids
(those whose rocks are less resistant than those of other
rapids), so that the pools which the rapids separated become confluent in short graded reaches. The longer surviving rapids are also gradually obliterated[19], so that the
graded reaches become longer and longer, and are eventually
reduced to grade with the lower graded course of the river.
This process is slowly extended upstream until only the
headwaters are torrential; and in time these are also
graded. The longer the upheaved landmass stands still, the
gentler the fall to which even the headwaters will be worn
down. An old river will have a graded course of very
gentle concavity and very faint fall all along its course.

The initial sag line, or depression followed by a consequent stream, will usually be more or less irregular, and
the stream will depart to one side or the other of a direct
course in bends of more or less pronounced curvature. The
gentler the initial fall and the more open the floor of the
sag, the more pronounced the bends are likely to be.

A small consequent stream, following such a course, will
flow around every bend, large or small, and the accidental
infall of trees or turf may develop new bends. By reason of
centrifugal force, which throws the thread of fastest current
toward the outer or concave bank, that bank will be cut
away while the opposite, inner or concave bank is built outward with detritus; thus although the stream does not gain
a greater breadth, the bends are all enlarged; and the stream
becomes notably serpentine in many bends of small size.

A larger river, beginning in a similarly bent course,
will obliterate many of the smaller bends, and will not
allow the formation of many new small bends by infalling
obstacles; but will enlarge the larger bends; hence its
course will come to have a small number of bends of larger
size. But it is to be noted that torrents of irregular
flow do not persistently enlarge their bends because they
do not persistently attack the outer or concave bank of the
bends. Bend enlargement is best accomplished by streams
that maintain a graded flow during and after uplift[20].

River dynamics

In a stream with an unobstructed channel the swiftest flow
will be near the center, well above the bottom, and slightly
below the surface. This is because there is a considerable
friction of the bed and banks, and also some friction with
the air above. The latter is most effective when the wind
is blowing upstream. In time of flood, a stream will increase its volume in two ways - first by a rise of the

surface, and second by a lowering of the bottom, and by
picking up the sediments on the bed (Fig 8).

During the rise of a river in flood its depth is doubly
increased, for as its surface is raised its bed is excavated.
The latter effect is due to increase in velocity and carrying
power, thereby much detritus is lifted from the river bottom
(said to be 30 or 40 feet in some rivers)[21]. As the flood
water is drained away the river slackens its flow, the
raised detritus is deposited again, but farther downstream.

Figure 8. Diagram to show manner
in which a stream gains volume
during flood periods: a to a' -
by rise in surface; b to b' - by
lowering of the channel bottom
(from a blackboard sketch).

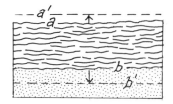

In early youth of a river, when it possesses a rushing
flow, its channel may for a time be beset by many rock
ledges, which divide the flow into a number of separate
streams (Fig 9). As such a stream rushes between its con-
fining ledges, its surface next downstream is higher than
the surface of the slower-moving water on either side,
sheltered by the two ledges. The rushing current therefore
sinks, and the water rises in boils on the smooth surface[22].
A good example of such a river is the Tchu River in Tibet.
It heads in Issuk-kul, a mountain lake, and falls for miles
in continuous rapids - a characteristic feature of the
early stages in a greatly uplifted region.

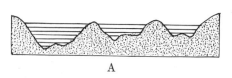

A

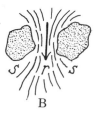

B

Figure 9. A rushing river: A) Cross-section, showing rock
ledges which divide the flow into separate streams.
B) Plan, showing rushing water r and stagnant water s
(from a blackboard sketch).

Although the surface of a swiftly-flowing stream can be
smooth, the friction on the bottom may be so great that the
surface is thrown into <u>standing waves</u> (Fig 10). In contrast
to ocean waves, these stand still while the water passes
through them. They may be so strong as to curve over and
break like surf - always upstream in the direction of wave
movement relative to the water. These waves produce rapid
scour and form large ripple-marks on the bed that move

Figure 10. Standing waves in a rushing stream (from a blackboard sketch).

gradually downstream[23]. Soundings indicate that the bed of the Mississippi River is thus ripple-marked on a grand scale.

The capacity of a stream to carry load is measured with respect to the velocity of a single thread of current, and increases with a high power of that velocity - said to be the sixth power, although this does not seem to hold good in practice[24]. The capacity of two rivers of unlike velocity can be compared by considering the number of threads of current[25] which each can apply to carrying their loads, and the velocities of the threads.

In a young, deep, narrow stream, threads that come in contact with the bottom are relatively few, and it is from here that most of the load is derived. Nevertheless, these threads are so active that they do practically all the load carrying. They quickly sweep away all the finer detritus and ask for more. The surface threads, running most rapidly, have no work to do. Such a young stream is clear, except when a rain floods it with turbid water, washed down from the valley sides. When such streams are in flood, they are rivers on edge (Fig 11A). As only the bottom threads are picking up and carrying material, they are not busy streams[26].

In a mature river, broad and shallow, the busy threads are much more numerous and are better placed to do work (Fig 11B). As there is much more detritus available, the river is turbid. When a young river becomes mature, its capacity is not changed as much as would be implied by its decrease in velocity. Its breadth is increased both by change in cross-section and by decreased velocity, with a consequent increase in the number of busy threads.

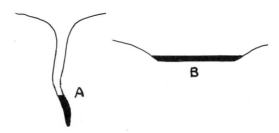

Figure 11. Cross-sections of (A) a cleft stream, and (B) a normal stream. In A, corrasion on the bottom greatly exceeds that on the sides, and only a few current threads are busy, whereas in B much more water is busy (from a blackboard sketch).

The thread of most rapid current, normally at mid-
stream a little below the surface, is displaced toward the
outer or concave bank at every bend. That bank is therefore
corraded and the bend becomes more pronounced; the belt of
bends widens also[27]. The detritus thus acquired is swept
along, and some of it is applied to building out of the
inner or concave bank on the bends farther downstream; thus
the river maintains a fairly constant width.

In consequence of this manifestation commonly known as
'centrifugal force', the river surface will be held a little
higher near the outer bank than near the inner bank at
every bend (Fig 12B). The water at the bottom, flowing
slower, will be pressed away by the higher surface water
(Fig 12C). The bottom flow thus systematically aids in
sweeping detritus from the undercut outer bank toward the
next downstream inner bank, where some of it may be deposited.

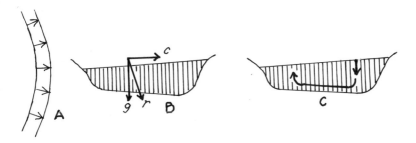

Figure 12. A) Plan of a river bend to show direction of
centrifugal force. B) Section of a river bend to show
resultant r of a centrifugal force c and gravity g.
The water surface will be at right angles to the
resultant, and thus higher on the outer bank.
C) Section of a river bend to show oblique bottom
flow (from a blackboard sketch).

Soils and soil creep

When a highland is reduced to a peneplain, the whole
surface, apart from belts of valley alluvium, is cloaked
with a mature, sedentary soil, from which all soluble
constituents (especially limestone) have been leached out.
At an advanced stage of leaching, even the silicates have
been largely decomposed, the silica has been dissolved and
carried away, and the bases, chiefly hydrated alumina, are
left, more or less stained with iron oxides. Such soils
may differ more according to the climate under which they
developed than according to the rocks from which they were
derived. In warm climates they are commonly reddish, and
are called laterites.

Different rocks weather in different ways. Granite on
slopes often yields large boulders between which a finer
residue is found. The same granite reduced to a peneplain
has a fine residual soil, gradually passing down to firm
rock at depths of 10, 20, or 30 feet. Slates may yield a

relatively fine soil, even on slopes. Limestones retain only a thin soil on slopes; on lowlands they are often cloaked with a red or yellow clay soil, representing the insoluble residue after the limestone has been dissolved away; the clays lie immediately on firm subsoil rock, of very irregular surface form.

The soils of an upheaved lowland are sooner or later swept away in a second cycle of erosion, and new soils are then slowly developed by the weathering of the lowered sur-face as it is denuded and degraded. The new soils on the subdued crests of hills on the divides and subdivides are purely of local origin, as they result from decomposition of the rock between them. They are commonly thin and stony when first developed; later, as the hills are worn down lower, the soils become finer textured and deeper. Soils on hillslopes resemble those on hilltops, except that they are more mixed by wash and creep. Soils on bottomlands of broad valley floors are alluvial, and contain material from all the drainage area above them comminuted, leached, and mixed.

Accumulated residual soils from the bedrock alternately expand and contract on slopes as a result of frost and thaw, heat and cold, and wetness and dryness. Expansion occurs normal to the slope, but contraction occurs vertically[28], thus causing a gradual downhill migration, or <u>creep</u> (Fig 13).

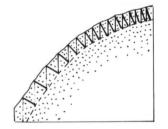

Figure 13. Soil creep produced by freezing and thawing. During freezing, soil expands normal to the slope. During thawing, it contracts perpendicularly (from a blackboard sketch).

Creep is aided by the growth of plant roots which push aside the soil; more is always pushed downhill than uphill (Fig 14). On the death of the plant the root decays, and the cavity is filled by soil that has crept from above. The same is also true of animal burrows. An example of the work of creep is the downhill bending of the surface of steeply dipping foliation or layering in schists (Fig 15) - a common feature in the southeastern States and elsewhere in regions of deep and prolonged weathering. Creep is most rapid on steep slopes[29], and weakens progressively downhill. Much detritus is delivered to stream sides by soil creep.

In wet weather, nearly all parts of a land surface may be covered by water rills, running down the lines of slope[30]. At all times there is also a very slow movement by creep of weathered rock fragments down the slope lines. Slope lines and movement along them may be classified in the same manner as rivers are classified: as well-defined, ill-defined, undefined - or as consequent, insequent, subsequent, obsequent, etc.

A river, when considered in the grossest manner, should be regarded as expanding all over its drainage area in wet

Figure 14. Soil creep produced by growth of tree
roots. A) During growth,
soil is forced downhill.
B) During decay, soil from
uphill moves into the
cavity (from a blackboard
sketch).

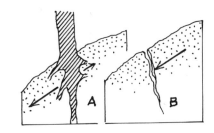

Figure 15. Downhill
bending of layering in
schist, as a result of
soil creep, (from a
blackboard sketch).

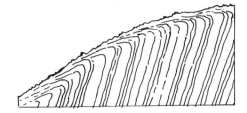

weather, and as shrinking into its channel in dry weather[31].
The lines of slow but persistent soil creep, as well as the
lines of rapid runoff, should be regarded as its
contributaries.

Ground water and river erosion

As a river deepens its valley, the ground water surface is
lowered toward it. Innumerable springs, small and large,
deliver the ground water to the river along its banks. The
slope of the ground water surface (inadvisably called the
'water table', although it is not level) will increase after
rain, by a rise of the interstream areas, because recharge
is more rapid than discharge; it will slowly weaken during
drought.
     Ground water very slowly moves through the interstices
of the soil and underlying rock in the direction of the
slope of the water table, and is thus led to sag-line
streams (or to the seashore) along the channels of which it
emerges in countless springs. Spring water is said to
'rise', but this rise always depends on the sinking of ground
water under the nearby higher ground. The movement of ground
water from its place of intake to its emergent springs is so
slow that it may remain underground for scores of years.
The spring that marks the head or source of a river gains
undeserved prominence. As valleys are deepened, the water
table is correspondingly lowered.
     During drought the slope of the water table is slowly
lowered; many springs shrink or fail, especially in the
upper valleys; the headwaters shorten; and the rivers are
reduced to a low stage of diminished discharge. After rain
provides a new supply of ground water, the slope of the water
table is steepened, the supply of springs is increased, the
headwaters are lengthened, and the rivers rise to a high stage

of increased discharge. In a large river, the difference
between low and high stages may be 40, 50, or 60 feet.

The work of a river is done chiefly during its great
floods[32]. At low water it plays in the channel determined
by the flood waters. A river system must be regarded as an
expanding and contracting entity. When snow lies on the
frozen ground and a spring rain falls, no water enters the
soil, all the rain and melted snow water runs off, and the
flooded river system is temporarily expanded to a leaf-
like sheet, covering the whole drainage area to the very
divides. In a dry season the leaf-system is skeletonized,
leaving only the stream ribs.

Aggradation and degradation

When a river is flowing as a torrent and cutting its valley
down to grade, it has greater carrying power than load[33].
It tends to cut a narrow, trench-like channel in the bedrock. As it cuts down, its fall, velocity, and carrying
power are all decreasing, while its load (especially that
part which comes from the valley sides) is increasing.
Thus, as the load is increasing and the carrying power is
decreasing, they come in time to be equal; then grade is
reached. But on reaching the grade and then ceasing to cut
downward, the channel is broadened by lateral erosion, and
as the channel is broadened the river depth is decreased.
But with decrease in depth, velocity is lessened by greater
action of friction. On flowing slower, carrying power is
weakened; and some of the increased load must be laid down
(all the coarsest detritus and some of the finer), thus
building up or aggrading the channel and increasing the
fall, the velocity, and the carrying power, so that the rest
of the load may be carried. The laid-down detritus is zero
at the river mouth, and increases in thickness upstream,
thus aiding in the extension of a graded course into the
torrent course; the detritus also laps on the valley sides
increasingly upstream and broadens the valley bottom[34].

If mature rivers receive much relatively coarse detritus,
they cannot lift it, but must now sweep it along the channel
bed; for that purpose they preserve a fairly strong fall.
Their banks are beset by shifting channels and lozenge-
shaped sand and gravel banks on which the river subdivides
into interlacing or braided channels. The sandbanks are
much modified by every flood. Such rivers are navigable, if
at all, only with great difficulty; but if not too large,
can be forded almost everywhere, by choosing a zigzag path.

If the load of a mature river is very fine, the fall
will be small, the river will lift much of it and will become very turbid; the channel will meander with slowly
changing curves. Such a river is easily navigable where
deep water flows around the outer side of a meander, but
with difficulty at the inflections or crossings, where
shifting sandbanks are formed. The river can only be forded
at such inflection points, and not at the concave bends.

During the aggradation thus caused, some of the river
water will be withdrawn from its visible current by per-
colating in the channel-bed detritus. As the river volume

is thus decreased, it will again be unable to carry all its
load; it must therefore continue to lay down some of its
load (the coarsest and part of the finer) in order again to
increase fall, velocity, and carrying power, so that the
rest of the load may be carried. Here again, aggradation is
zero at the mouth and increases upstream. The river thus
becomes underfit[35].

In the meantime, the dissection of the highlands is
advancing and the area of steep valley-side slopes is increasing; hence the load brought down to the main river
valley by its branches will increase also. But with this
increase in load, the river must again lay down some of the
load and steepen its fall, in order to carry the rest.

The graded river will usually increase its length slowly
by enlarging its bends. But with increase of length, fall
will be decreased. To prevent that, aggradation must again
be resorted to. In all these possibilities, aggradation is
zero at the mouth, and increases upstream, hence it is
divergent aggradation.

Detritus brought down to the river mouth will form a
forward-growing delta, if the marine agencies there in
action do not prevent it by sweeping the detritus off into
deeper water. If a delta is formed, the river course is
lengthened; and in order for the river to carry its load
across the delta, the whole of its graded length must be
slightly aggraded by laying down a little of its load; but
in this case the aggradation is of almost uniform depth all
the way up its graded course. This is parallel aggradation,
in contrast to the previously discussed possibilities of
divergent aggradation.

A time will come, as maturity passes, when the valley-
side slopes are reduced to less and less declivity; then the
load that they furnish to the streams will be decreased and
refined, and the aggraded river will become a degrading
river; for with decreasing load its carrying power will be
in excess, and to prevent such an excess it will take up
some of the previously deposited detritus. It thereby at
once decreases its fall, velocity, and carrying power, and
increases its load until a balance is again brought about.
This process does not go by fits and starts, except insofar
as floods are concerned; but is a continuously acting process, always maintaining carrying power and load in equilibrium[36].

As maturity advances into old age and river load is not
only decreased in quantity but refined in texture, all the
aggraded valleys may be degraded to fainter and fainter
fall; so that when old age is entered upon the rivers will
be again working on bedrock, all the aggraded detritus
having been swept away[37]. This result will be reached
sooner if marine agencies cut away the coast and shorten
the rivers, for then also they must degrade their whole
length - or just the opposite of the case in which aggradation of the whole length was caused by increase of river
length through delta growth.

The erosion cycle

Meander development

Ordinarily, the radius of curvature of initial bends is greater than half the distance between points of inflection; hence, the arc of the bends is less than 180°. As a result of corrasion later on, the curvature of such bends is usually sharpened and the radius of curvature shortened, the arc becoming a semi-circle (Fig 16A)[38]. The tangent-limited belt of such enlarged curves is wider than the initial belt of the bends. After this, both the radius and the arc of curvature become larger (Fig 16B)[39]. When the curves are well developed and of roughly similar size, they become meanders[40].

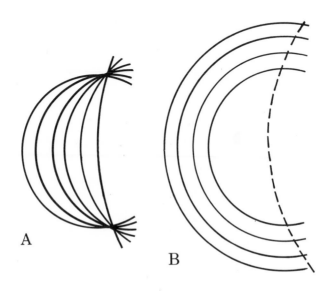

Figure 16. Diagrams showing changes in curvature of river bends: A) Early stage, in which curvature of bends is increased and the radius decreased. B) Later stage, when curvature remains constant, and the radius increases (from a blackboard sketch).

The initial bends of a consequent river, more or less modified and enlarged while the river is degrading its original course, are more freely developed on floodplains than before. The thread of fastest current, deflecting toward the outer bank, cuts the bank away, but it does not begin this cutting at the beginning of the bank-curve, and it continues this cutting along the same bank beyond the end of the curve (Fig 17). This results in the downstream shift of lateral erosion (Fig 18). The river thus enlarges its curves, widens the meander belt, increases the angle between its path and the axis of the meanders, and shifts all the meanders down-valley[41].

Bedrock meanders may form in youth of the first cycle by lateral corrasion on consequent bends. On the outside of

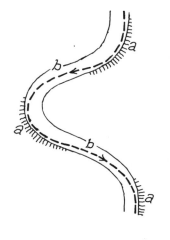

Figure 17. Diagram showing how lateral cutting extends downstream, past the point of inflection, resulting in down-valley migration of river bends. a) Point of cutting. b) Point of crossing (from a blackboard sketch).

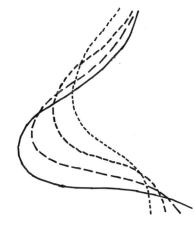

Figure 18. Diagram showing successive forms assumed by river bends during their downstream migration (from a blackboard sketch).

each bend, and continuing downstream past the point of in-flection, the river will undercut, exposing much bare rock. If the rock were firm enough, the undercut slope would over-hang, but it rarely does so, and the overhang breaks down into a steep wall. Its form suggests the term amphitheatral wall[42] (Figs 19 and 20).

As the river bends are continually migrating downstream and outward, and as the river is downcutting at the same time, the inside of each bend will have an outward and down-stream slope, gentler than that of the amphitheatral wall, and more or less cloaked by river gravel and silt. The manner in which this is formed by the streams suggests the term slipoff slope[43].

For a considerable time there will remain on the up-stream part of each curve a spur of unconsumed upland, pro-jecting into the valley. Thus, the rim or the upland out-lines a valley of winding pattern - an incised meandering valley (Fig 20). The river, flowing in a smooth curve

19

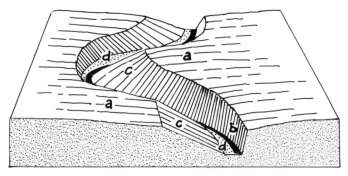

Figure 19. Block diagram of an incised meandering valley.
   a) Undissected upland. b) Amphitheatral wall. c) Slipoff slope. d) Floodplain scroll (from a blackboard
   sketch.

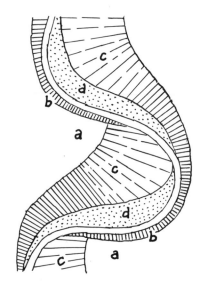

Figure 20. Map of an incised
   meandering valley. Letters
   are the same as those on
   Figure 19 (from a blackboard
   sketch).

around the base of the concave amphitheater, is <u>competent</u>,
or <u>fit</u>[44].
     When such a river reaches maturity it will continue its
outward and downward migration, but will cease downcutting.
Thus, on the rounded ends and downvalley sides of the spurs
the river will build out detrital deposits. These deposits
lie on a corraded rock floor, the coarsest at the bottom,
the finest at the top. The top of the deposits will be at
river flood level. The surface of the deposit has the
pattern of a narrow scroll, of which one cusp points upstream around a spur end, and the other downstream into an
amphitheater. The scroll is the beginning of a valley-floor
floodplain, and can be termed a <u>floodplain scroll</u> (Fig 20).
The area of a floodplain scroll increases by accretion.

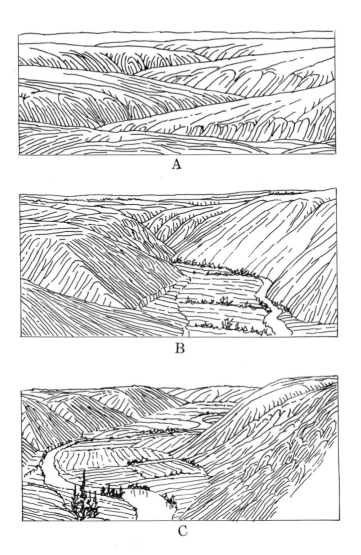

Figure 21. Processes of valley deepening and valley
widening illustrated along the River Tardes, a small
head stream of the Loire drainage basin south of
Montlucon, on the north side of the Central Plateau of
France. A) Incised meanders in the upper course,
looking downstream. B) A tributary with a gorge above
and a narrow valley flat below. C) The open valley of
the lower course (from Davis, 1912, Fig 51, 52, 53).

When an observer on top of one of the spur tops looks upstream in an incised meandering valley, only the slipoff spur slopes are visible; as these are commonly cleared and cultivated, the valley seems well occupied. When he looks downstream he sees only the steeply undercut amphitheatral spur sides, and the valley appears to be wild and deserted (Fig 21A).

The direction of flow of a stream in an incised meandering valley can be readily determined from the map pattern, as the amphitheatral walls consistently extend downstream from the point of inflection of the bends (Fig 20).

If the valley depth is greater than the radius of the stream curve, as may happen along small streams, the adjacent amphitheaters will become confluent at moderate heights above the stream, and above that height the spurs are consumed (Fig 22). Although the baseline of each amphitheater follows the large arc of each stream curve, the upper rim of each amphitheater will have only a small arc of larger radius and gentler curvature. The valley, outlined by such a rim, will not be nearly as sinuous or meandering as the stream. Such a valley has incised meanders at the bottom, but they have little effect on the gross pattern of the valley.

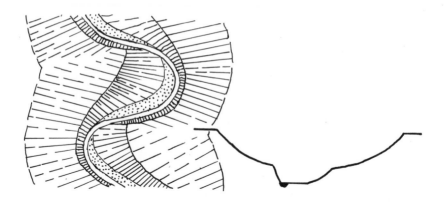

Figure 22. Map and section, showing pattern of upland rim of a small incised meandering valley, where adjacent amphitheaters become confluent, thus giving the valley a less sinuous course than those in Figures 20 and 21 (from blackboard sketch).

In consequence of the down-valley migration of the whole system of river curves or meanders, the undercut walls of the amphitheaters on the upvalley sides of the spurs will persist, although with decreasing height. Amphitheatral and spur forms thus become less pronounced as lateral and vertical corrasion advance together. Spurs are narrowed, shortened, blunted, and consumed, and the valley becomes an open valley (Fig 21C).

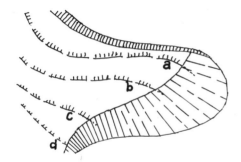

Figure 23. Map showing
    successive trimming of
    spur ends: a) Narrowed.
    b) Sharpened. c) Shortened.
    d) Blunted - and consumed
    (the last stage not being
    shown) (from blackboard
    sketch).

    As the spurs are trimmed, narrowed, etc., the floodplain
scrolls will be widened (Fig 24a). By the time the spurs
are consumed, the first-formed part of each scroll will have
been swept away, and the widened patch of floodplain in a
river meander will have become <u>pouch-shaped</u> (Fig 24b).
The valley floor is then about as wide as the meander belt.
Nevertheless, the successive pouch patches are still unconsumed.
    After the valley-side spurs have been consumed and the
river has an open valley floor to work on, its pouch-
patches are constricted at the necks to form lobes
(Fig 24c).
    Further growth of the meanders will produce cutoffs
(Fig 25). The abandoned meander loop will become an oxbow
lake, with its ends filled by river silt; the lake will
gradually be converted into a swamp, and in time smoothed

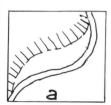 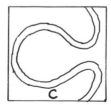 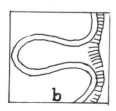 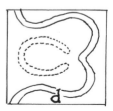

Figure 24. Terminology of floodplain meanders: a) Scroll.
    b) Pouch. c) Lobe. d) Later stage, unnamed (from a
    blackboard sketch).

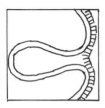 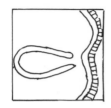 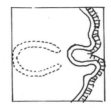 

Figure 25. Stages of floodplain development. After a
    meander is cut off, a new meander forms a short distance
    below the former one (from a blackboard sketch).

23

over. In the meantime, the abrupt short-cutting of the
river will cause the curvature of its channel to enlarge
downstream, producing a new meander a little below the cutoff. Other minor modifications of the channel are produced
upstream and downstream from the cutoff[45].

The sudden change of a river course from a roundabout
meander loop (or to an overflow cut-through) is legally
termed <u>avulsion</u>. When rivers form boundaries the boundary
remains in its original position if the courses are changed
by avulsion, but the boundaries change with the channel if
the change is by corrasion and <u>accretion</u>[46]. Oxbow lakes
mark the maximum width to which meanders can grow, and the
tangents drawn to such meanders mark the maximum meander-
belt width. This width is seldom greater than 18 times the
width of the river channel[47].

The surface of the widening floodplain scrolls is built
up most rapidly close to the river bank where, at times of
overflow, the first check in velocity occurs, and where the
most silt is laid down. Hence, a floodplain slopes away
from its river, as well as downvalley. As the growth and
downvalley shift of meanders is not uniform, being most
rapid in times of flood, the growth of floodplain scrolls
takes place by starts and shifts, and its surface is faintly
belted parallel to its growing side.

A side stream, entering the main valley, will find that
a direct course to the main river is a little uphill,
because the floodplain slopes away from the <u>natural levees</u>
along the main river. The side stream must therefore turn
downvalley, and run along the base of the slipoff slope to
the end of the scroll, where the river will receive it
(Fig 26). This part of the side stream is commonly
marshy[48].

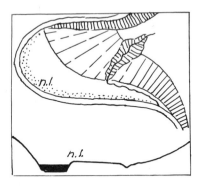

Figure 26. Map and section
  showing the effect of
  natural levees (n.l.) on a
  tributary entering a floodplain (from a blackboard
  sketch).

A river that has developed floodplain scrolls, and has
also aggraded its valley floor because of the load received
by the side streams, may therefore lose so much of its surface flow to the underflow that it is no longer competent
to sweep around its curves in a vigorous fashion. It may
then depart more or less from its fit path, and become an
<u>underfit river</u> (Fig 27). Similar changes can also be pro-
duced climatically, by reduction in rainfall[49].

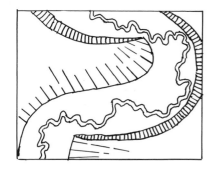

Figure 27.   Map of an underfit
river (from a blackboard
sketch).

A winding river generally increases its length by enlarging its curves. This decreases its slope and its
velocity, resulting in aggradation of the channel and
floodplain. However, increase in length by enlargement of
curves also increases the likelihood of cutoffs. Hence,
the length of a winding river slowly increases, then
suddenly decreases, thus oscillating above and below its
average length[50].
    When a cutoff occurs and the oxbow lake silts up, three
lobes of the floodplain become confluent (Fig 25). Because
of the cutoffs and other accidents, the meander belt will
not widen with perfect regularity, but more here than
there; in the course of time, the belt itself shifts
laterally right and left and the valley sides are undercut,
the valley itself will become wider than the meander belt.
Then, adjacent floodplain lobes are connected by a narrow
strip along the base of the valley-side bluff from which
the meander belt is withdrawn, while on the other side of
the valley the meanders are trimming off the valley-side
bluffs, after the fashion of a planing machine. Now and
then a meander belt will cut too far into the bluff, and
thus create a little amphitheater.
    The more continuous the bluff-base floodplain strip
which unites the floodplain lobes, the farther downstream
must a tributary flow before it can join the main river.
A remarkable example is the Yazoo River, which flows 150
miles along the edge of the alluvial valley of the
Mississippi River, before it can join the master stream at
Vicksburg.

Waterfalls

When a river flows across a succession of hard and weak
beds, it grades a segment of its course across each weak
belt with respect to a local baselevel determined by the
rock sill at the next downstream hard bed (except that a
weak belt next to the shore will be graded with respect to
the baselevel of the ocean surface). Such a graded segment
is a reach[51]. As a hard-rock sill is slowly worn down, the
graded reach next upstream is correspondingly degraded;
thus the river flows evenly from one weak belt to the next
hard belt. But in passing from a hard belt to a weak belt
the river will make a sudden descent, providing the

succeeding hard belts stand lower and lower down-valley. At the sudden descent is a <u>cascade</u> (if the change from hard to weak rocks is gradual) or a <u>fall</u> (if the change is abrupt). If a hard-rock belt is somewhat broad, it will have torrential <u>rapids</u> between its upper limit, where the next graded reach joins it, and the lower limit, where the cascade or fall occurs.

A river will be of greatest breadth in the soft-rock graded reaches where it runs slowest; it will run faster through the hard-rock rapids, and there its foaming current will be narrowed; it will be slender at the falls. Abrupt falls will excavate a <u>cave of the winds</u> behind them and a <u>plunge pool</u>, often of considerable breadth and depth beneath them. Fallen blocks of the hard fall-maker and fragments of the weak rocks will be churned about in the pool, especially during flood.

Let the hard beds vary in hardness. The less hard ones will be worn down faster than the hardest ones; their falls will diminish in height and in time will be extinguished, whereupon two graded reaches will become confluent across the extinguished fall-maker (Fig 28). At the same time, a fall from a harder bed next upstream will increase in height to a maximum, then slowly diminish in height to disappearance. Eventually all the falls will be worn down to grade with respect to baselevel, and the river will gain a normally graded and concave profile all along its length[52].

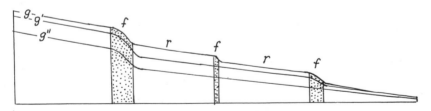

Figure 28. Profiles showing stages in erosion of a sequence of resistant and non-resistant strata. r) Graded reaches on non-resistant beds. f) Falls on resistant beds. g, g', g") Successive gradients of stream (from a blackboard sketch).

If the hard belts are determined by resistant strata, the inclination of the strata will influence the duration of the falls. Falls on strata dipping steeply upstream will be extinguished much sooner than on level strata, because the downwearing of level strata to grade (as deter-mined by the upstream prolongation of the graded river below the falls until it intersects the fall-maker) is more work than the back-wearing of the steep strata to grade (Fig 29). If the fall-making stratum is steeply inclined, the valley will be constricted where it is cut through; its outcrops can commonly be traced from the banks of the rapids up the valley side (Fig 30A). If the fall-making stratum lies level, it will be continued more or less

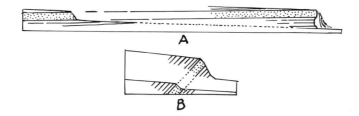

Figure 29. Profiles showing how waterfalls are obliterated
more slowly on gently dipping strata (A) than on
steeply dipping strata (B) (from a blackboard sketch).

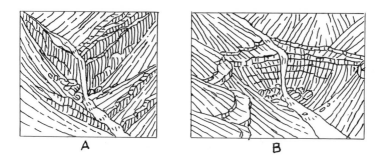

Figure 30. Sketches of waterfalls: A) On vertical strata.
B) On horizontal strata (from Davis, 1908, Fig 3C,
pl. 22).

visibly in cliff-like ledges along the sides of the
widening valley downstream from the falls; but the weak
underlying strata, which form slopes from the cliff-ledges
down to the valley bottom, are excavated into a vertical-
faced cave of the winds back of the falls (Fig 30B).
    A good example of the contrast between steeply dipping
and horizontal strata occurs in the Appalachian Mountains
and Allegheny Plateau in Pennsylvania. Thick, resistant
beds which are steeply inclined cross the Susquehanna River
above Harrisburg, but have been so worn down that they produce only the faintest of riffles. In the plateau to the
west, by contrast, much thinner and less resistant beds
which lie horizontally still form notable waterfalls.

Subsequent drainage and river capture

Imagine two young consequent rivers - one, c, large, the
other c, small - flowing in about parallel courses across
a region in which the strata dip steeply upstream; these
strata include two wide belts of resistant rocks, separated
by a narrow belt of weak strata (Fig 31A).
    As the rivers are cut down - c cutting down faster and
deeper than c - they will widen slowly in the resistant

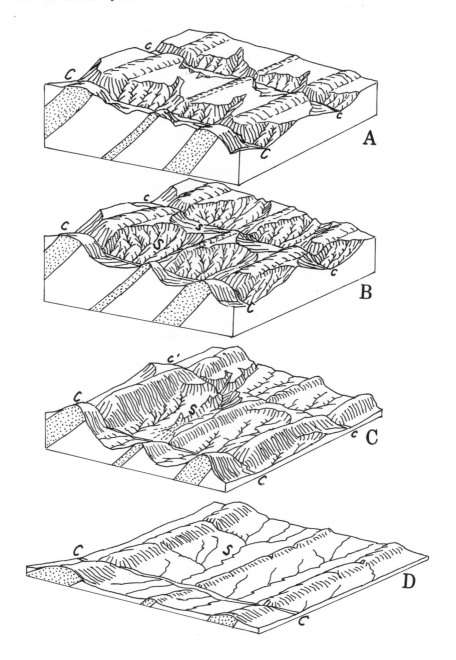

Figure 31. Block diagrams showing development of sub-
    sequent streams, with a river capture in block C.
    s and s are subsequent streams, which are branches of
    consequent streams c and c (from Davis, 1908, fig 2, 4,
    6, 8, pl. 21, 22, 24).

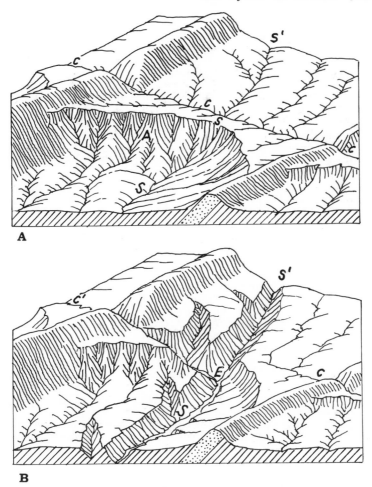

A

B

Figure 32.   Diagrams showing stages of river capture.   In
     the upper diagram (A), capture is imminent, and in the
     lower (B), it has occurred in the near past.
     s) Subsequent stream, draining to a large consequent
     (not shown, but see Figure 31).   *c*) Small consequent.
     s) Subsequent draining into *c*.   *c'*) Beheaded part of *c*
     (in lower diagram).   *A*) Alcove.   *E*) Elbow of capture
     (from Davis, 1912, fig 89, 90).

belts and rapidly in the weak belts.   The young valleys
will be rocky defiles in the resistant belts, but will
have a lateral expansion, or alcove (*A* in Fig 32) on each
side in the weak belts.   A small branch stream will run
down each lateral expansion, as the surface will be worn
down lower than the groundwater surface.
     These little streams will grow longer by retrogressive
wash or headward erosion, following the guidance of the
weak belts.   Such streams are <u>subsequents</u>.   Usually they
join the consequents in pairs.   Consequent valleys are very

commonly symmetrical in cross-section (except for bends
and meanders) and have rocks and forms of the same kind
facing each other, right and left, across the valley.
Subsequent valleys, however, are commonly asymmetrical in
cross-section, and have unlike rocks and forms facing each
other, right and left.

Examine the opposite-flowing subsequents, between the
two consequents c and c. One of them, s, joins the larger
consequent c at a lower level than that at which the other,
s, joins the smaller consequent c (Fig 31B). For a time,
the initial upland surface between the two retrograding
heads of s and s will not be wholly consumed; drainage of
this upland will be vaguely defined. But later on, the two
subsequent valley heads will meet in a sharp dividing edge.
When this stage is reached, s will be longer than s, and
(if the distance between c and c is not excessive) the
divide will have a steeper slope, and will therefore shift,
slowly losing height the while, toward c. s will gain,
s will lose length and volume.

This shift, or migration of the divide, will continue
until it comes close to c. Groundwater from below c will
by that time leak through to the head of s, and accelerate
its further retrogressive lengthening. Finally, the upper
part of c will be captured and diverted to s, leaving the
lower part of c beheaded (Fig 31C). Upper c will thereupon
cut a gorge around the sharp turn at the point of diversion
(the elbow of capture, E of Fig 32), and will deepen its
valley upstream therefrom, because it has been given a
lower gradient to the ocean, via s and c. s is thereby
enlarged and becomes an overfit stream; it enlarges its
curves and degrades its course; lower c is enlarged and
becomes somewhat overfit, although not as much as s. It
may, despite its volume, have to aggrade its valley floor
somewhat, because of the active downwash of detritus from
the upper part of c and s. Beheaded lower c is diminished
and becomes underfit - strikingly so near the elbow of
capture. Its uppermost section may be reversed into s,
thus shortening it still further.

The beheaded river is left with so small a volume that
it is unable to keep the former channel clear of detritus
washed in by side streams, especially for a few miles below
its head; the channel is clogged, and the little stream
takes whatever course it can on the aggraded valley floor.
If the valley were originally of a serpentine pattern, in
which the river flowed in smooth curves, it can do so no
longer; it now staggers irregularly around the curves as an
underfit river. Near its head, the detritus brought in by
a side stream may accumulate in a small alluvial fan, which
will grow across the course of the weak beheaded stream,
and compel it to rise in a swamp or pond upstream from the
fan. The rise may be sufficient to reverse the flow of the
ponded stretch, and cause it to flow backward into the
gorge at the elbow of capture. The divide between the
two river systems is then shifted from the edge of the
gorge to the obstructing fan.

While s and s are in opposition, the divide between
them slowly creeps toward c (Fig 32); when the capture of
upper c is made, the divide suddenly leaps from the elbow

and lassoes all the upper drainage of *c*. Whatever headward erosion was previously in progress by *c* will be accelerated as the deepening of *c* around the elbow is propagated upstream.

The subsequent branch *s'* of *c*, which is opposite to *s*, will be captured at about the same time as upper *c*, and will therefore deepen its valley (Fig 32). *s* will thus gain in length and *ss'* may extend its backward erosion to capture still another opposing subsequent – provided that this subsequent does not drain down steeply to another large consequent. Thus, the area tributary to *c* is further extended. Hence the tendency is for large consequents, running transversely across belted hard and soft structures, to grow still larger by the aid of their predatory subsequents, at the expense of the smaller consequents.

Instead of only three belts of resistant and weak strata, as above, we may have many successions of resistant and weak belts, transverse to the courses of *c* and *c*. The smaller consequent *c* will then be subdivided into many parts, each captured at many points by many small subsequent branches of *c*. Drainage of the district will thus come to be largely by subsequents along the weak belts; only the master consequents will persist in transverse courses across resistant and weak belts alike. Falls and rapids, at the passages from resistant to weak belts, are thus greatly reduced in number. Drainage is then said to be, on the whole, adjusted to the structures of the weak-rock belts. At the same time, the uplands and ridges between the valleys are adjusted to the resistant structures. The abandoned notches across the ridges, formerly water-gaps through which the smaller consequents formerly flowed, now survive only as wind-gaps. Good examples of both water-gaps and wind-gaps occur in the north-central Appalachian Mountains, where a history much like that which has been outlined seems to have taken place (Fig 33).

Six successive stages of progress before and after river capture may be described as far-future, approaching, imminent, recent, well-established, and long-past captures[53]: the first, when an unsymmetrical divide suggests that a capture may eventually take place, although much erosion must be done before it is accomplished; the third, when the head of a degrading branch is close to, and is reinforced by springs leaking from a high-perched river; the fourth, when a narrow gorge has been cut by a locally-steepened torrent around the elbow of capture; the sixth, when indications of capture are largely removed by the adjustments of the rivers to the resulting drainage area.

An approaching capture is indicated by the relation of the high-perched Kanawha (= New) River in North Carolina (flowing northward to the Ohio) to certain streams that flow by lower-cut courses southeastward to the Atlantic and that are actively cutting back the steep slope of the unsymmetrical Blue Ridge divide at their heads; the divide to the New River being relatively short and gentle[54]. A recent capture is indicated by the sharp-cut valley around an apparent elbow of capture where the western angle of South Carolina nudges the eastern boundary of Georgia; the capturing river is the Savannah; the beheaded river is the Chattahoochee, flowing southwestward to the Gulf of Mexico.

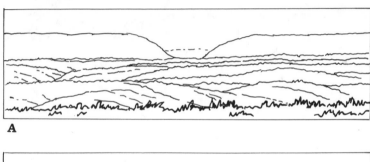

A

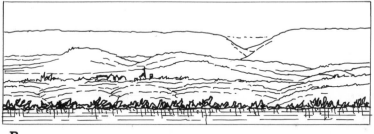

B

Figure 33. Sketches of Appalachian water-gaps and wind-
gaps. A) Delaware Water-gap, on boundary between
Pennsylvania and New Jersey. B) Wind-gap near
Susquehanna River in Pennsylvania (from Davis, 1898,
fig 14, 18).

The Meuse, flowing north-northwest from France into
Belgium, has been delayed in deepening its middle and upper
course because its middle course traverses the hard-rock
mass of the highlands of the Ardennes in a gorge. Its
upper course formerly received a long branch from the
southeast and a shorter one from the south. Nearby tributaries of the Rhine on the east and the Seine on the west
are deeper-set than the high-perched Meuse. Several of
their branches grew headward toward the Meuse branches; a
successful Rhine branch captured the southeastern branch of
the Meuse, and a successful Seine branch captured the
southern branch of the Meuse. Thus the Meuse is left with
a slender drainage basin between its neighbors. The Meuse
is distinctly an underfit river, for it staggers around
its valley curves; but it is underfit upstream from the
junction of the lost branches as well as downstream, where
its underfitness must have some further cause than loss of
volume by capture.
Drainage can also be captured by another process than
that just discussed. When two rivers happen to curve
toward each other, they will slowly gnaw into the ridge
between them and they may in time consume it. Thereupon
the smaller river occupying the less deep valley (already
somewhat diminished by leakage of groundwater to the deeper
river) will drop by a low fall into the other river. A
capture of this sort is by <u>lateral abstraction</u>[56]. The fall
at the point of capture will gradually be worn back, and
the smaller river will in time be graded to its new trunk

river. The remainder of the smaller river, downstream from
the point of abstraction, is diminished in volume, and
becomes a beheaded, underfit river. It may be somewhat
shortened by the reversal of its upper part back of an
obstruction.

If the smaller river is a branch of the larger one, the
point of confluence will be shifted upriver when lateral
abstraction occurs; and the group of hills between the
former and the new point of confluence may be called inter-
confluence hills. Crowleys Ridge in the Mississippi
alluvial valley is a good example[57].

The Arid Cycle

A desert is any region barren of life,[58] and of such
regions there are many kinds, depending on reasons for
their barrenness – frozen deserts, fresh lava flows, bare
rock surfaces produced by degradation,[59] and regions that
are arid (hence very dry). The latter concern us here, and
we will consider the processes of degradation that proceed
according to the arid cycle.

In terms of landforms, an arid region is one whose
rainfall is insufficient to fill the original structural
basins,[60] and to send drainage to the sea. Its water-
courses are withering streams[61] – raging torrents during a
few days of the year, and dry channels during the remainder.

The ideal arid cycle

In formulating the arid cycle, we shall assume an ideal
case: An extensive region of deformed structure, worn down
to low relief, is uplifted, upwarped, or faulted with
sufficient rapidity to extinguish most of the previous
drainage. The climate of the region is dry enough so that
the streams flow intermittently, and so that most or all of
its rivers and streams are unable to flow to the sea.

Early stages

Drainage from the deformed terrain will be prevailingly
consequent on the new structural surface, and will gather
into centripetal groups flowing into separate enclosed
basins – each a disconnected ganglion (Fig 34A). Erosional
processes will begin even during the deformation, but they
will be much enhanced during the succeeding and longer
period of stillstand.

As desert mountains have scanty vegetation, detritus is
actively swept from their slopes when occasional heavy

34

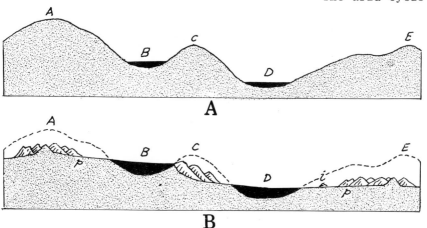

Figure 34. Sections showing early stages of an ideal arid
    cycle: A) Immediately after the initial deformation,
    with consequent divides at *A*, *C*, and *E*, and consequent
    basins at *B* and *D*. B) A more advanced stage, after
    much filling of the basins and reduction of the divides.
    Pediments *p* have developed at the margins of the
    basins; the mountain fronts have steepened during
    recession, and there are a few inselbergen *i*; basin *B*
    is about to be captured by basin *D* (from a blackboard
    sketch).

rains fall; the fans outside the mountain valleys are
therefore of relatively strong fall, contrasting with the
fans of faint fall formed of finer detritus by larger
rivers of wet-weather mountains[62]. The intramont basin
floor will therefore be aggraded to a <u>slowly rising base-
level</u> in each basin (Fig 34B). The central parts of the
basins will contain <u>playas</u> or <u>salt flats</u>, which will shift
from place to place as detritus is spread toward them from
one direction or another. The playas are salty because
they retain all the salt brought in by the rivers. When
the flats are covered by water - either from rainfall
directly on their surfaces, or from wash from surrounding
areas - the salt will swell and heave, producing a rough,
jagged ground that is very difficult to cross over. The
salty ground, being hygroscopic, will hold moisture, and
will resist removal by the wind.
    The streams of each basin will degrade the surrounding
highlands, producing the detritus which fills the central
part. Moreover, for reasons that will be explained later,
the mountain fronts will stand steeply, yet will retreat
by the work of gravity, weathering, and rill wash. As they
do so, they will leave behind them belts of bare <u>rock
floors</u>, or <u>pediments</u>, which separate them from the detrital
area in the center of the basin[63].
    The largest intramont desert basin in the world lies in
central Asia. Lop Nor is a shallow lake on its flat floor;
the lake is so shallow and the floor is so flat that a
flood, washing in sediment from the mountains on one side,

drives the lake away when the sediments are deposited in a
thin delta; the lake shifts its place and outline. Many of
the insloping fans at the mountain-base margin of this
basin are now moderately trenched by the rivers that built
them, as if by a change in climate. The plateau of Tibet
contains many smaller desert basins.

Integration of basins

Because of the different altitudes of the various drainage
basins, they will be separated by unsymmetrical divides.
On an asymmetrical divide between a higher and a lower
basin, the fall will be greatest toward the latter, and its
more vigorous streams will shift the divide toward the
higher basin (Figs 35, 36). In the higher basin, large
fans may form on the side away from the lower basin, but
only small fans will form on the nearer side, where the
drainage was smaller originally, and which is being further
subdivided by erosion of streams draining into the lower
basin (Fig 36B,C). Eventually these streams will push back
the divide into the higher basin, capturing its drainage
area (Fig 36D). Its deposits, and even the bedrock floor
beneath them, will be ripped by the streams draining into
the lower basin.

Figure 35. Profiles
   showing progressive
   shifting of divide
   between two desert
   basins in process of
   integration (from a
   blackboard sketch).

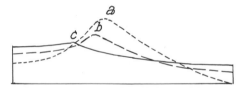

This process results in integration of the originally
separated structural basins, and is a distinctive feature
of the arid cycle[64]. No matter how many structural basins
there may have been immediately after the initial deformation, the largest and deepest basin of all will in time
acquire all the others.

An asymmetrical divide may be seen west of Bonneville
Flats, the western part of the Salt Lake Basin of Utah,
where the Western Pacific Railroad winds up the east-
sloping ravined mountains, to emerge on a higher, still
undissected basin in Nevada; eventually this will be
captured by the Salt Lake Basin. The late Tertiary deposits of Dripping Springs Valley in Arizona (shown in the
Ray Folio) are being ripped and removed by the Gila River,
which has captured the 'valley' after a period when it had
a closed, high-level basin.

Probably the best place to find basins in the process
of integration would be in the Tibetan Plateau of Central
Asia, but this region is little known. However, on its
south side desert basins are now being encroached by
streams that drain into deep gorges in the Himalaya
Mountains. Some of the headwater torrents of the Himalayan
gorges have already gnawed their way through the northern-

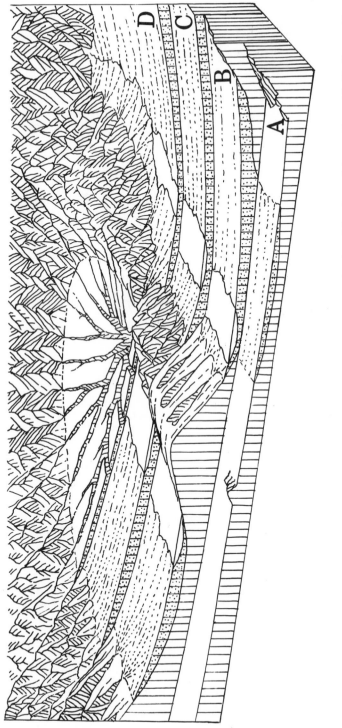

Figure 36. Diagrams showing the early and middle stages of the ideal arid cycle: A) Before the deformation that initiated the cycle. B) After the initial deformation, and after moderate erosion and deposition. C) A more advanced stage in the development of the independent basins. D) Integration of the basins, with capture of the higher basin by the lower; deposits of the higher basin, and the floor beneath them, are being ripped by streams draining into the lower basin (from Davis, 1912, fig 142).

Figure 37. A high-level, interior basin at the southern
    edge of the arid Tibetan Plateau, on point of capture
    by steeply-plunging streams tributary to the low-lying
    Himalayan drainage (from Davis, 1898, fig 200).

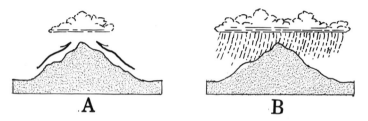

Figure 38. The effect of mountains in increasing the rain-
    fall in an arid region. A) Convection currents cause
    condensation of local rain clouds above the mountains.
    B) During cyclonic storms, rain falls on the mountains,
    but evaporates before reaching the lowlands (from a
    blackboard sketch).

most ranges, and capture of the basins beyond is imminent
(Fig 37). The notches which they have cut through the
mountains form passes by which the Tibetan Plateau is
reached from India to the south.

Results of integration

Once a single lower basin has captured many or all of the
higher basins, its rising baselevel will control the whole
region. Detritus will pour into it from all the captured
higher basins, and will aggrade most of its surface.
Farther back, rock floors and pediments will be extended
toward the remaining mountainous areas, whose steeply
sloping faces will retreat until they are consumed.
    Now, there is a pronounced increase in aridity. During
earlier stages, although the rainfall of the region is
scanty, the mountainous areas produced convection currents
which formed local storm clouds over their summits, from

which rain could fall (Fig 38A). Moreover, when cyclonic
storms crossed the region, their rain, which might evaporate
before it reached the lowlands, was able to descend on the
mountains (Fig 38B). When the area and height of the
mountains is reduced by erosion, these opportunities for
rainfall become less and less common.

Late stages of the cycle

During early stages of the arid cycle, dominant work of
erosion is by streams, even though they are ephemeral, and
work of the wind is ineffective. Winds may sweep over the
mountains, but these mountains will shield the basin floors
from their attack. Moreover, what erosion is performed by
the wind will be negated by the more effective work of the
streams; any depressions eroded in the lowlands by the wind
will be filled in by detritus washed from the surrounding
mountains by the streams.

Progress of the cycle conspires to enhance the work of
the wind and to reduce the work of the streams. Reduction
of the mountain areas decreases the rainfall and the effec-
tiveness of the streams; it also diminishes the volume of
the detritus which is being transported into the basins,
and the detritus still being transported is of finer texture than earlier. The extent of the lowland areas increases, both by broadening of the depositional floor and
by extension of rock floors around it, which, coupled with
the reduction of the mountain areas, gives the wind a clear
sweep across vast areas. Moreover, the increasingly finer
texture of the detritus now supplies the wind with a
potentially greater load for transportation.

Nevertheless, the flow of the winds differs from the
flow of streams in being generally planless and fickle, so
that their erosional results are often negative. Winds may
swish the dust and sand about, carrying it from place to
place, or even returning it to its starting point. If the
winds have a prevailing direction, much dust may be carried
by them, until it settles on a plant-covered surface, where
it cannot be picked up again and carried farther (or it may
be carried to the sea, as where trade winds sweep sandy
dust from the Sahara into the adjoining Atlantic). The
settling dust tends to accumulate on an even surface, and
to decrease the inequality of the district where it gathers,
forming a plain. Its accumulation is so uniform that it
does not exhibit stratification, except in marginal areas,
where thin intercalated layers of inwashed gravel and sand
may be formed. Deposits of dust are known by the German
name of loess.

Many dust-filled basins, now more or less dissected,
are known in the interior of China. (Probably a drier
climate prevailed during their accumulation than now).
These dust deposits are fertile and support a large popu-
lation; dwellings are frequently excavated in the loess of
the valley sides. Although not indurated, the walls and
roofs are enduring, but they collapse during earthquakes.

Processes of advanced aridity will result in leveling
without baseleveling. The surface of the desert will be
lowered by removal of material from the region by the wind,

yet wherever the wind excavates a local basin, stream flow will wash in detritus and level it.

The whole region will thus wear down, lower and lower, without regard for ocean level, its only limit being the persistence of the same climatic conditions. Conceivably, an interior basin could be cut below sealevel by wind excavation and associated arid processes. Such a basin, if extended, might eventually tap the sea, causing a sudden flooding by water - a possibility worthy of consideration by paleogeographers[65].

Recapitulation

In this ideal cycle, two special conditions have been assumed to bring it to completion: (a) a perfect stillstand of the crust at the beginning of the cycle, and (b) a stationary climate, except insofar as erosional reduction of the region increases its aridity. These conditions are seldom, if ever realized[66]. Possibly they approach nearest to realization in the Saharan and Arabian Deserts, parts of which contain lowlands whose extent greatly exceeds that of the uplands and mountains, which are virtually waterless, rainless, and streamless, and where the wind seems to be the only effective agent at work, sweeping across wide surfaces of bare rock, and building up the sands into enormous areas of dunes.

Most other arid regions, such as those in the western United States, do not nearly approach such conditions - mountains form a sizeable proportion of their areas, some vegetative cover exists, even in the lowlands, and stream work, although scanty, still dominates the erosional regime. In such regions, crustal disturbances (which were partly responsible for the aridity itself) were geologically rather recent, and even still persist in places (see discussion of the Basin Ranges, below), so that a prolonged stillstand has not been attained. Moreover, the aridity of most regions is likely to be interrupted by external circumstances before the final stages of the cycle are attained. A geologically momentary interruption in aridity by external cicumstances occurred during the ice ages of Pleistocene time, which were accompanied by marked increases in rainfall in certain areas. During these times, the previously arid Great Basin of the western United States was extensively flooded by Lakes Bonneville, Lahontan, and others, some of which briefly drained out to the sea.

Details of arid landforms

Lowland areas

During early stages of the arid cycle, lowland areas will be created by filling of structural depressions with detritus derived from erosion of surrounding highlands, producing level plains with an aggradational surface. As the cycle progresses and the highlands are worn back, more and more of the lowland areas will be degradational rather than aggradational, and these will finally dominate altogether.

Figure 39. Pediment compartments, or incipient pediments, at the base of the western faulted front of the granitic and gneissic Santa Catalina Mountains north of Tucson, Arizona (from Davis, 1931, fig 4).

A

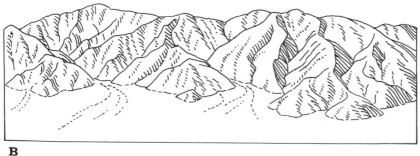

B

Figure 40. Fan-bayed mountains, which have receded some distance from their original structural fronts, the intervening area being covered by the upper parts of the fans, which lie on an eroded surface of the bedrock (sub-alluvial bench of Lawson, see below). Presence of bedrock near the surface in the upper parts of the fans is suggested by projecting mountain spurs, and by the detached bouldery mount in the upper view. In a slightly later stage, the inner edge of the eroded surface of the bedrock will emerge as a pediment along the mountain base. Both views are in the Mojave Desert region of southeastern California. A) North face of Silver Mountain near Victorville. B) West face of mountains of southern extension of Death Valley trough. (from Davis, 1938, fig 20, 22).

A

B

Figure 41. The angular junction between slopes of granitic
   mountains and the pediments at their bases (A) compared
   with the curved junction in non-granitic mountains (B).
   Both views are in the Mojave Desert region of southeastern California. A) South face of Granite Mountains
   northeast of Barstow, looking east. B) North base of
   Ord Mountain east of Victorville, looking east (from
   Davis 1933, fig 2 and Davis 1938, fig 13).

These degradational surfaces are <u>rock floors</u>, or
<u>pediments</u> - smooth surfaces of bare rock, thinly veneered
by detritus in transport across them, which conditions
their form and gradient. They begin to form in a belt
between the central aggradational area of a basin and the
mountains along its periphery, at a time when these mountains have reached an advanced stage of erosion, and have
retreated considerably from their original structural
fronts (Figs 39 and 40).
   The term 'pediment' implies a surface which slopes away
in all directions from a central monument - a still unconsumed mountain mass. Many bare-rock surfaces surround such

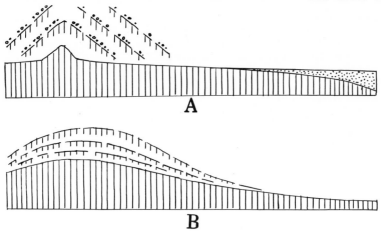

Figure 42. Comparison between slope profiles in arid regions A, and in humid regions, B (from a blackboard sketch).

monumental mountain masses, but many others have no parti-cular relation to any mountain, so that the term 'rock floor' seems more appropriate for the whole genus. Never-theless, in most publications the term 'pediment' seems to have preempted all the possible variants - such expressions have been used as 'pedimentation' for leveling under arid conditions, and 'pediplain' for the form produced.

Highland areas

Under arid conditions, the still unconsumed mountains behind the pediments form slopes whose angle is determined by the nature of their bedrock, and the weathered rock derived therefrom - rocks which weather to large spalls (such as granites) form steep slopes, whereas rocks which weather to smaller spalls (such as sediments and volcanics) form gentler slopes (Fig 41). The angles of these slopes result from a grade adjusted to the materials in transport across them, by the processes of gravity, weathering, slopewash, etc. The same angles are maintained in each type of rock during retreat of the highland areas, so long as the masses of particular rocks are unconsumed[67]. This condition contrasts with that in the humid cycle, in which the angle of slope in rocks of all kinds is progressively lowered with time (Fig 42). Regardless of the variations in angle of slope of the mountain fronts, all of them rise more steeply than the pediments below them - abrupt in the more massive (granitic) rocks, and rounded in the less massive (sedimentary and volcanic) rocks (Fig 41).

The abrupt junction between mountains and pediments, formed late in the arid cycle, has been confused in many regions with abrupt junctions resulting from tectonic causes earlier in the arid cycle - such as the fault-scarp fronts of the block mountains in the Basin Ranges, discussed below.

As erosion advances and the mountain fronts retreat at a constant angle, the mountain mass will be reduced to smaller and smaller areas, and these will be broken into separate groups, and even isolated knobs (Fig 43). For the latter Bornhardt (1900), whose work was in Southwest Africa, proposed the term inselberg (= island mountain). These 'islands' differ from true islands that project from the sea, in that the surrounding pediments are not level, and in that they are carved from the same material as the inselbergen themselves[68]. Thus, the name applied to these 'island mountains' might better have been monticules - were it not that 'inselberg' has acquired wide acceptance in the literature.

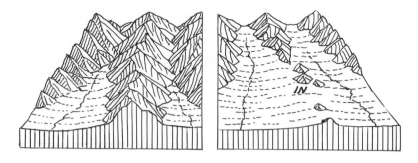

Figure 43. The formation of inselberg *IN* by spreading of pediment tentacles, and retreat of intervening mountain spurs (from a blackboard sketch).

An advanced stage of degradation will be the pan-fan - a dome-like area of degraded rock from which the central monument has disappeared, except for a few inselberg at most. The term was proposed by A.C. Lawson (1915) and is somewhat inappropriate, inasmuch as the surface of the dome is degradational rather than aggradational, and is mantled thinly, if at all, by fan material; the awkward word 'pan-pediment' might have been more accurate. Many such domes occur in the Mojave Desert region of southeastern California (Figs 44 to 47) - one of the finest, and the one originally cited by Lawson, being north of Cima on the Union Pacific Railroad, in the Ivanpah Quadrangle. This Cima Dome, and others in the same region, are 5 to 10 miles in diameter and rise as much as 2000 feet above their bases. Presumably, now that their central mountains have disappeared, they will gradually be lowered and will disappear in the late stages of the cycle[69].

In humid regions, upland slopes fade downward into lowland slopes, and are not separated from each other by an abrupt angle (Fig 42). Arid regions resemble these humid regions in that their unconsumed highlands stand above stream grade and have yet to be reduced by stream work. In the upland areas of both regions, weathering loosens rocks and slopewash transports them, with the aid of gravity. In the upland areas of both, weathered materials move slowly

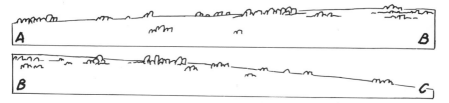

Figure 44. Panorama of a granite dome, or pan-fan, from
   which all mountains have disappeared, leaving only
   clusters of residual boulders. The Noble Dome north of
   Barstow in the Mojave Desert region of southeastern
   California, looking east (from Davis, 1938, fig 30).

Figure 45. Residual boulders on granite domes, or 'jumping
   jacks' (so-called by Davis "from their exceptionally
   alert appearance"). Views are in the Mojave Desert
   region of southeastern California. A) In Rand
   Mountains near Randsburg. B) On Noble Dome north of
   Barstow (from Davis, 1938, figs 21 and 31).

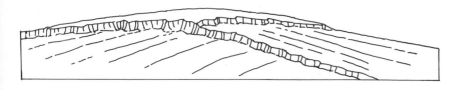

Figure 46. A granite dome, of pan-fan, with discordant
   surfaces that drain into several playas, each with a
   different baselevel (from Davis, 1938, fig 12).

A

B

C

Figure 47. Unsymmetrical divides on granite domes, or
    pan-fans, produced by the conditions illustrated
    theoretically in Fig 46. All views are in the Mojave
    Desert region of southeastern California. A) Manchester
    Divide northwest of Needles, looking northeast.
    B) The same, looking south. C) Bullion Mountains
    southwest of Cadiz, looking north (from Davis, 1933,
    figs 17, 27, and 28).

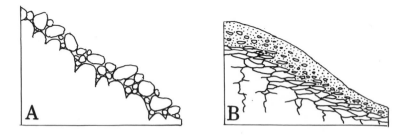

A

B

Figure 48. Comparison between weathered rock surfaces in
    arid regions A, and in humid regions B (from a black-
    board sketch).

but steadily downslope. Nevertheless, in arid regions, the whole surface is rocky, without soil or residuum, so that creep is negligible (Fig 48A). In humid regions, residuum and fine-grained soil accumulate on the weathered surfaces, partly with the aid of a considerable mantle of vegetation, hence movement of the weathered material is greatly aided by creep (Fig 48B).

In arid regions, weathering and its products thus do not mask the essential change in transporting processes on proceeding from the highland to the lowland areas – from crude processes at work on highland slopes (gravity, weathering, and rillwash), to the more refined processes at work in the adjacent lowlands (sheetwash and stream-wash). Not only is there a change in process but a change in texture of materials; when materials on mountains slopes have marched to their bases, they have weathered down into fragments fine enough to be transported by streams and sheetfloods across the pediments.

Fluctuations in equilibrium of pediments

Most observed pediments of the southwestern United States do not form smooth surfaces in the process of creation today. Instead, they are either thinly veneered by detritus or have been ravined and dissected to shallow depths. These disturbed pediments epitomize the very delicate balance under which they were created – a balance between the nature of the materials, the competence of the transporting agencies, the tectonic environment, and the general climate (as between more arid and more humid). Even variations of these features within small limits will bring about disruption of the theoretically smooth pediment surfaces in the manner commonly seen today[70].

Historical sketch

Recognition of results of leveling by degradation in arid regions came slowly, and it was widely assumed for years, especially in the United States, that the lowlands of the desert were dominantly or wholly aggradational. Thus, C.F. Tolman (1909) described the bajada, or aggradational detrital slope, as the characteristic landform of the desert, without apparently realizing its special nature, as caused by filling of a basin by detritus of a geologically recent tectonic depression, or that filling of a basin by detritus would ultimately and inevitably be supplanted by erosional leveling.

In the United States, much of this misunderstanding arose from entangling the pediment problem with the Basin Range problem – two related but quite different matters. Any statement to the effect that lowlands in the Basin Ranges were degradational rather than aggradational was considered to be an attack on the validity of the block-fault origin of the Basin Ranges – and in fact was so used by opponents of the block-fault concept.

The degradational nature of surfaces in desert regions of prolonged erosion was actually observed early (1897) by W.J. McGee, in the Sonoran Desert of northwestern Mexico,

where he found to his astonishment that his wagon was
rolling over worn-down surfaces of weathered granite, miles
from any mountain area. He ascribed the leveling of these
surfaces (probably with a large measure of correctness) to
sheetflood erosion.

These able pioneer observations were long either
ignored or not believed, and recognition was hindered by
the numerous writings of C.R. Keyes (1909, 1912), an
opponent of the fault-block concept of the Basin Ranges,
who asserted that most of the lowlands in the Basin and
Range province were degradational, and had been leveled,
not by water, but by the wind. His extreme, and often
mendacious statements brought discredit to the whole concept.

Thus, much of the early recognition of the true nature
of desert forms came from German geomorphologists working
in Southwest Africa - although their explanations of these
forms seem contrived and implausible. Bornhardt here
observed inselbergs but explained them by a complex process
of uplift and subsidence, and by undercutting of criss-
crossing streams. Passarge (1904) attempted to explain the
forms of the region by an alternation of dry and humid
climates.

Following McGee, next valid recognition of degradational surfaces in the American desert was by Sidney Paige
(1912) and by A.C. Lawson (1915), although they regarded
them as minor and incidental features in a region of
dominant alluviation. Paige ascribed them to undercutting
of streams issuing from adjacent mountains. Lawson, in his
'Epigene profiles of the desert', reasoned that, as the
mountain fronts retreated and as the detrital plains rose
and advanced upon them, a degraded suballuvial bench would
form beneath the deposits which, in the later stages, would
emerge near the mountain front as a subaerial platform
(Fig 49).

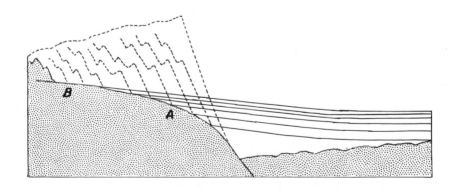

Figure 49. Development of a suballuvial bench A and sub-
    aerial platform B, as postulated by Lawson, from an
    original structural surface (assumed to have been a
    fault block), as a result of retreat of mountain slopes
    and encroachment of alluvium in a region of slowly
    rising baselevel (from Davis, 1930a, fig 2).

Complete formulation of the concept of desert degradation was made at last by Kirk Bryan (1922) as an incidental
product of an investigation for the U.S. Geological Survey
of routes to desert watering places in southwestern
Arizona; his 'Erosion and sedimentation in the Papago
Country' is now classic. He correctly described the many
desert landforms of the region and the processes at work
on them, and demonstrated the wide extent of rock-cut
surfaces around the mountain areas (hitherto only dimly
visualized and unnamed). For them, he proposed the term
pediment, now generally accepted as the title for this
whole class of landforms - "the desert-inhabiting species
of the genus peneplain" in the words of Eliot Blackwelder
(1931).

PART 2 COMPLICATIONS OF THE

EROSION CYCLE

Simple conditions were assumed in the preceding discussion
of the cycle of normal erosion - namely, a soil-covered
plain of degradation produced on a uniform rock mass at
the end of a former (or first) cycle of erosion by long-
continued action of weather and rivers; uplifted with
moderate warping, and then standing still indefinitely
through a second cycle of erosion. These simple conditions
must now be modified.

The rock structure of the landmass need not be uniform;
it may be varied in many ways (massive crystalline rocks,
horizontal stratified rocks, or tilted, folded, broken or
faulted stratified rocks, etc.).

The landmass need not have been reduced to a plain of
degradation with ill-defined river systems when the uplift
that introduced the second cycle of erosion took place; it
may have been a peneplain, a region of subdued mounts and
hills, or a maturely mountainous region, with a well-
defined river system; and if so, parts of all these river
systems would be carried over from the first into the
second cycle of erosion.

The upheaval by which the second cycle is introduced
may have been disorderly, with irregular breaks (faults),
flexures, folds, etc. New movements of irregular upheaval
or subsidence may occur at any time, interrupting the
second cycle at any stage of its progress, introducing a
third cycle.

We shall now continue with a series of selected
examples, relating mainly to differences in structure.

Plains

Coastal plains

A coastal plain is formed of a body of marine strata that
have been laid bare by emergence, due either to uplift or
lowering of sealevel. Characteristically, its strata slope
gently seaward, partly as a result of original seaward dip,
partly because of tilting during uplift or downbending of
the coast during sedimentation. The coastal plain is
bordered inland by an oldland, formed of the basement of
the coastal plain deposits, composed of earlier rocks of
varied and perhaps complex structure. On these, the
coastal plain deposits lie unconformably, and against them
the deposits overlap inland, toward an original shoreline.

First-cycle coastal plains

A first-cycle coastal plain is bordered inland by the
original shoreline at the edge of the hilly oldland, and
seaward by a new shoreline resulting from emergence
(Fig 50).
 In youth, the surface of the first-cycle coastal plain
is the surface of its uppermost stratum - the last to have
been deposited before the emergence. On this is a sheet of
soils derived from this stratum, 'which is uniform throughout, except for seaward variations in texture, away from
the original shoreline. The plain is drained by consequent
streams which extend down the structural and depositional
slope, some originating in the oldland, some in the coastal
plain itself.
 As maturity approaches, the original width of the
coastal plain is reduced on both its landward and seaward
sides (Fig 51). Seaward, marine erosion attacks and re-
moves the edge of the coastal plain deposits. Landward,
the original marginal deposits of the coastal plain are
stripped from the edge of the oldland, laying bare the worn

51

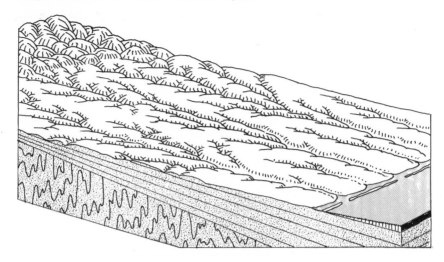

Figure 50.   A young first-cycle coastal plain (from
    Davis, 1898, fig 75).

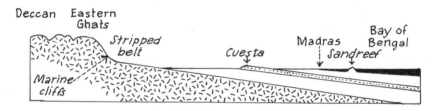

Figure 51.   Late mature first-cycle coastal plain on south-
    eastern coast of India (from a blackboard sketch).

down surface of the basement on which they had been
deposited.   In this stripped belt, the oldland rocks have
a gentle seaward slope of low relief, contrasting with the
original hilly oldland farther inland, which was never
buried by deposits.   In the stripped belt, consequent
streams that originally flowed from the oldland onto the
coastal plain have been <u>superimposed</u> on the oldland surface.
    Within the coastal plain itself, approaching maturity
is marked by increasing dominance of the larger consequents, which now cut broad valleys below the level of
the original coastal plain surface.   This original surface
remains in the interfluves as broad flat uplands, which
may be termed <u>doabs</u>, from the Hindu word meaning 'two
rivers'.   During maturity the doabs become dissected by
branch <u>insequent streams</u>.   As downcutting progresses, the
original upper stratum of the coastal plain is breached,
and lower strata are exposed.   These lower strata may include one or more units of greater resistance than the
rest, which are carved into <u>cuestas</u>.
    With old age, the width of the coastal plain is still
further reduced, by stripping of coastal plain deposits

from the oldland on the inland side, and by marine erosion
on the seaward side. The doabs and cuestas are worn down
to new lowlands, in which the outcropping edges of the
sloping underlying rocks are expressed less by relief
features than by belts of contrasting soils and vegetation.
    Examples of a young first-cycle coastal plain occur on
the west coast of Scotland, and on the south side of the
Gulf of St. Lawrence in Quebec. A late mature coastal
plain occurs in the Madras district of southeastern India
(Fig 51), where the Deccan Plateau forms the oldland and
is bordered by cliffs produced by marine erosion. Seaward
from these cliffs is a broad, gently-sloping stripped belt,
underlain by oldland rocks. Within the coastal plain
itself is a retreating cuesta of small height. The sea has
only moderately eroded the edge of the plain, and has
occupied itself mainly in building up offshore sand reefs.

Second-cycle coastal plains

The second cycle in coastal plain evolution may be
initiated by renewed uplift of the area, bringing to view
an additional belt of coastal plain deposits along the seaward side. As a possible variant, a period of subsidence
may intervene between the two uplifts, drowning the shoreline and embaying the lower ends of the consequent
valleys (Fig 52).
    In the second cycle, the coastal plain contains two
elements - the rejuvenated older coastal plain toward the
interior, and the newly exposed young coastal plain toward
the coast (Fig 53). During the second cycle, accentuation
of cuestas will proceed apace. Now, there are not only
belts of soils, but also a belted topography, consisting of

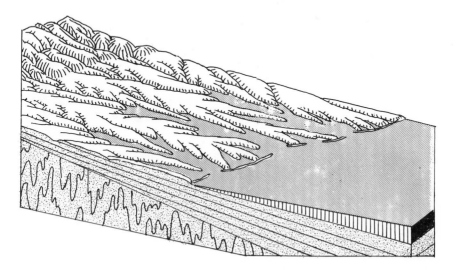

Figure 52. A coastal plain, after the first cycle has been
    terminated by a moderate submergence (from Davis,
    1898, fig 80).

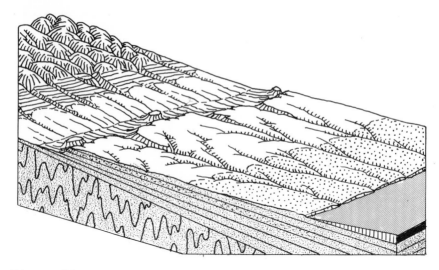

Figure 53. A second-cycle coastal plain. Renewed uplift has exposed a younger coastal plain seaward from the older. In the older one, erosional adjustments have produced a stripped belt at the edge of the oldland, an inner lowland, and a cuesta (from Davis, 1898, fig 82).

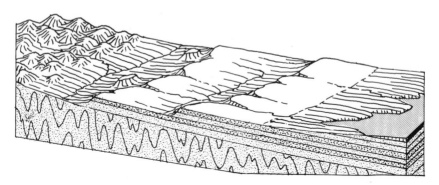

Figure 54. A coastal plain in a plural cycle, with prominent cuestas, and subsequent drainage (from Davis, 1898, fig 87).

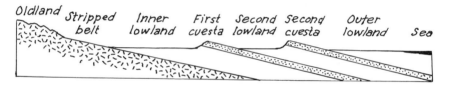

Figure 55. Terminology of coastal plain topography (from a blackboard sketch).

lines of cuestas, separated by belts of lowland. As
erosion progresses, the cuestas will move downdip, or seaward (Fig 54).

Coastal plain topography is now subdivisible into successive units, proceeding seaward (Fig 55): 1) The original
hilly oldland. 2) The stripped belt of older rocks.
3) An inner lowland. 4) The first cuesta. 5) A second
lowland, a second cuesta, and possibly others. 6) An outer
lowland, including the newly uplifted young coastal plain
along the coast. As there are many possible variations in
the nature and thickness of the strong and weak units in
the coastal plain deposits, a wide variety of cuesta and
lowland forms is possible (Fig 56). The cuestas may be
strong or weak, close-set to overlapping, or widely spaced.
With advancing age, the cuestas may be maturely dissected,
degraded, or faint.

 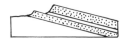 

Figure 56. Possible variations in cuesta forms (from a
blackboard sketch).

During the second cycle, competition between drainage
is strong. Subsequent streams grow along the lowland
belts, and frequently capture and divert the smaller consequents (Fig 56). Obsequent streams drain updip into the
subsequent streams on the escarpment faces of the cuestas[71].

Artesian wells

Coastal plains afford a good opportunity for drilling of
artesian wells, if their gently slanting strata include
some pervious layers (aquifers). The pervious layers receive rainfall at their outcropping surface and conduct it
underground down their slant along the aquifers. A well,
drilled down to the aquifer in the lower part of the plain,
where the surface is less elevated than the surface of the
aquifer outcrop, will give discharge to the water of the
aquifer, which may rise like a fountain above the surface
of the well. Artesian wells drilled in Maryland east of
Chesapeake Bay deliver water from aquifers whose intake is
west of the bay; the impervious aquifuges, overlying the
aquifers, keep out the salt water. Similarly, Atlantic
City, built on an offshore sand reef separated from the
mainland by a salt-water lagoon, is supplied with artesian
well water from aquifer intakes many miles inland in New
Jersey.

Atlantic and Gulf coastal plains

The Atlantic Coastal Plain in the United States illustrates
a coastal plain now in its plural cycle.

A pale cuesta begins in southern New Jersey and
strengthens northeastward. Its northeastern end, over-

looking New York Harbor, is the Highlands of Navesink. It
is continued, more or less covered by moraines, through
Long Island, Marthas Vineyard, and Nantucket, and is trace-
able by soundings across the Gulf of Maine. Its inner lowland, adjoining the backland, is in part drowned in the
upper northeastern part of Delaware Bay; it continues
northeast from the southwestern turn of the Delaware across
New Jersey to New York Harbor, to the production of which
its submergence contributes, and is then submerged in Long
Island Sound.

In Maryland, Virginia, and North Carolina, there are no
cuestas, but there has been an elaborate insequent dissection of the interior uplands, confluent with that of the
stripped lowland of the Piedmont Plateau. Since the
emergence and dissection of this extensive coastal plain,
its northeastern part, from North Carolina to New England,
has been increasingly submerged and thus has been narrowed
and embayed (Pamlico and Albemarle Sounds, Chesapeake and
Delaware Bays), until it ends at the Hudson embayment of
New York Harbor, thus causing the sea to overflow the
border of the New England backland. In the southern part
of the segment are extensive offshore sandreefs. This segment is now at least in its third cycle - two cycles resulting from uplift, and a third from submergence.

In South Carolina and Georgia there are neither cuestas
nor embayments; lack of embayments is because this segment
subsided less than farther north. In Alabama there is a
broad and much dissected cuesta of low relief, the
Cunnemugga Ridge, maintained by an infertile sandstone,
which runs east and west. Its inner lowland has a rich
black soil formed by decomposition of weak limestone; it is
known as the 'black belt' from the color of its soil. The
Gulf Coastal Plain has its greatest width in Texas, where
there are several low cuestas which lie inland from a very
young, low, smooth plain along the Gulf margin. The Texas
part of the coastal plain is in its plural cycle, with
moderate drowning along the coast.

Southeastern England

Southeastern England is a complex, plural-cycle coastal
plain[72]. The central and northern part of the country is a
hilly or sub-mountainous region, broadening in the north,
narrowing to the southwest. During the long-continued submergence of the surrounding area, the sea floor received a
heavy series of varied stratified deposits. Emergence
toward the southeast revealed a considerable area of sea
floor, whereupon the sub-mountainous region became a backland of the coastal plain (Fig 57). The southeastern
coastal plain is reduced to a belted plain of degradation,
in which two resistant formations are underlain, separated,
and overlain by weaker formations (Mesozoic) (Fig 58).

The region thus degraded was upheaved again, thereby
adding a new coastal plain to the southeast (Tertiary).
The weak formations are worn down to second-cycle lowlands,
leaving the resistant formations standing up as an inner
limestone cuesta (Jurassic) and an outer chalk cuesta
(Cretaceous). The inner lowland, worn down on weak red

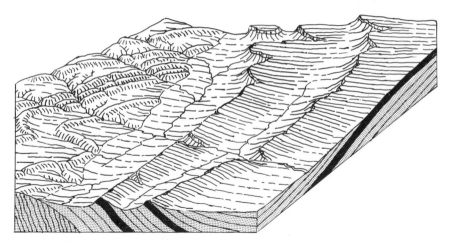

Figure 57. Contact between coastal plain rocks with a
   homoclinal structure, and oldland rocks of complex
   structure, as in southeastern England (from Davis,
   1912, fig 97).

sandstones and shales (Triassic) is connected by lowland
arms through the tripartite backland with a similar lowland
or coastal plain on the west.
   The northern backland includes Northumberland,
Yorkshire, Lancashire, etc; the middle backland is Wales,
and the southwestern one is Cornwall and Devon. The
drainage, having had a two-cycle opportunity of development, is well adjusted to the structures, and includes
master consequent rivers, subsequent branches, etc.
   The northeastern part of the two-cuesta coastal plain
is bent down eastward and submerged, so that a new north-
south shoreline is formed which is obliquely transverse to
the whole series of formations from backland to new coastal
plain (mostly resubmerged in the downbending) in the south.
Similarly, the southwestern part is bent down to the south
and submerged, so that a new east-west shoreline is formed,
which likewise traverses all the formations. A domed
structure (the Weald) occupies the southeastern angle
between the two new shorelines, and complicates this part
of the region.
   In consequence of the downbending and submergence, the
new shoreline enters more or less upon the lowland belts in
open embayments, and advances around capes and headlands,
where the cuestas dip into the sea. Beginning on the
northeast, the backland to the north has a ragged shoreline, cut back into cliffs by the sea; it contains coal
beds, extensively mined. The lowland, worn down on weak
red sandstones and shales, is a little invaded by an open
bay; here a small river, the Tees, enters the sea. A
little way north is Newcastle-on-Tyne, whence a great amount
of coal is shipped to London. The inner limestone cuesta,
though cut back in high cliffs, still stands forth as a
coastal salient (Whitby Cliffs). Next follows an open bay,

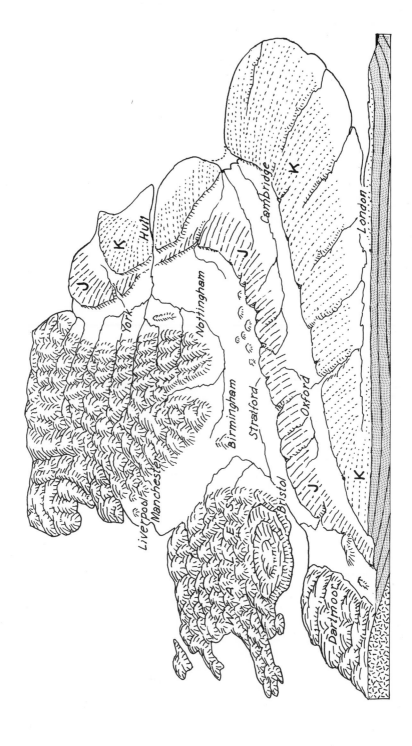

Figure 58. Southeastern England, an example of a complex, plural-cycle coastal plain area. J) Jurassic limestone cuesta. K) Cretaceous chalk cuesta (from Davis 1912, fig 99).

where the sea enters the inter-cuesta lowland (Vale of
Pickering). Then comes a sharper-cliffed headland, where
the outer (chalk) cuesta stands forth in Flamborough Head.
Farther south, the new shoreline obliquely traverses the
widening backslope of the outer cuesta and the outer lowland (Norfolk coast); here, Yarmouth, noted for its
fisheries, lies at the mouth of the Yare. Between
Flamborough Head and the Norfolk coast, this shore is,
about midlength, embayed by the shallow submergence of a
widened consequent valley in the outer cuesta; this part of
the embayment is The Wash, which is alternately bare at low
tide and covered at high tide; next inland come the marshy
Fens. Here a small river, the Cam, comes from the inter-
cuesta clay-lowland on the southwest. South of the Norfolk
coast, an embayment enters a trough between the coastal
plain and the southeastern dome; the inland, non-submerged
part of the trough is followed by the consequent, east-
flowing Thames. London is situated at the bay head. The
pre-eminence of this great city was established when the
southeastern coastal plain, the agricultural half of
England, was the seat of its greatest wealth.

On the south coast, the new shoreline is ragged along
the coast of Cornwall and Devon; here Falmouth is at the
narrowly embayed mouth of the Fal; Plymouth lies farther
east on the more broadly embayed mouth of the Ply; and
Exmouth is at the mouth of the Exe, which drains the
southernmost part of the inner lowland. Exeter is a little
inland on the lowland, which is here relatively narrow.
The inner lowland is continuous from Teesmouth to Exmouth.
It broadens through its midlength - the Midlands - where
its two arms pass westward through the tripartite backland
to the corresponding inner lowland of a western coastal
plain, nearly drowned by the Irish Sea.

The northern part of the inner lowland is drained by a
master consequent, the Humber, through the two cuestas to
the North Sea. On the narrowly embayed lower part of this
river is the important commercial city of Hull. The Humber
receives a long subsequent branch, the Trent, from the
southwest and a shorter subsequent from the north. The
divide between the latter and the Tees is largely determined by a glacial moraine; nearby stands the historic city
of York on the inner lowland. The Midland area of the
inner lowland, partly drained northeastward by the Trent,
is also drained to the Liverpool embayment on the west,
between Lancashire and Wales, by the Mersey and the Dee;
and also by the Severn southwestward to the Bristol Channel
embayment between Wales and Devon-Cornwall. A coal-
containing island of the backland, once completely buried
by the coastal plain strata, has been partly resurrected by
erosion in the Midland section of the inner lowland; there
stands the important manufacturing city of Birmingham.

The Severn (which heads in Wales) and its northeastern
branch, the Stratford (Avon), appear to have had some
advantage over the original consequent drainage of the
coastal plain from the Wales backland to the southwest[73];
for the headwaters of that drainage are now all diverted
to the Bristol Channel, and the beheaded lower waters,
constituting the Thames system, rise on the backslope of

the inner cuesta (there known as the Cotswold Hills) and
unite in a single consequent trunk - the Thames - on the
inter-cuesta clay-lowland; Oxford lies there. The trunk
river cuts its valley (Goring Gap) through the outer cuesta
(the Chiltern Hills), and then turns down the London
trough; Reading lies at the turn.

The two cuestas may, like the inner lowland, be traced
all across England, from the North Sea to the south coast.
They exhibit various peculiar features. Not far southwest
of Flamborough Head, inland from which the chalk cuesta
curves around from westward to southward, the weak strata
between the two cuesta-makers thin out and the two cuestas
unite in a single one. The cross-valley of the Humber is
cut a little to the south. There, as the clays thicken
again, the inter-cuesta lowland is again developed; it is
the Cambridge-Oxford lowland, named for the two university
cities located upon it. Through a middle length, the limestones of the inner cuesta have a gentler dip than to the
north and southwest, hence the cuesta there loses definition, and becomes ragged, with outliers detached from it on
the inner lowland. In about the same mid-length, the outer
(chalk) cuesta loses strength and is lower and weaker than
elsewhere. Farther southwest, where the inner cuesta
curves southward toward the south coast, it is interrupted
by a former island of the backland, which was completely
buried by the strata of the original coastal plain, but
which now, partly stripped of its cover, stands forth as
the Mendip Hills.

Paris Basin

In the Paris Basin of northern France, cuestas are an
important element of the topography (Fig 59). The Paris
Basin is, however, not a true coastal plain; dips are
centripetal toward Paris at the center, rather than in one
direction toward the sea. Moreover, the basin has had a
more complex history than the usual coastal plain; rivers
no longer pursue consequent courses across it; there have
been many erosion cycles, as well as much tilting and
warping since the emergence of the cuesta-making rocks.

On the east side of Paris are five east-facing cuestas,
dipping away from the Ardennes and Vosges oldland, which
are natural lines of defense in surface warfare. Southwestward and westward, these play out, so that only one
cuesta faces the oldland of Brittany on the west.

Other plains

Besides coastal plains, many other kinds exist - for
example, lacustrine plains, fluviatile plains, intramont
plains, aeolian plains, glacial plains, till plains, and
lava plains, all formed by surface accumulation of
materials, as well as different varieties of degradational
plains. Here, we will consider only two examples, derived
from the Great Plains region east of the Rocky Mountains in
western United States.

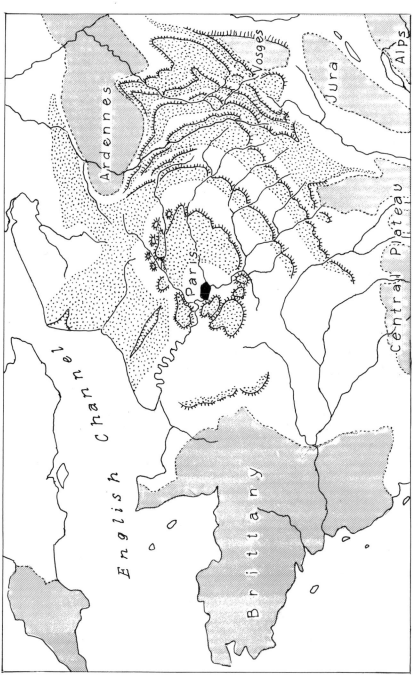

Figure 59. The Paris Basin of northern France, showing concentric cuestas, especially toward the east and northeast (compiled from various Davis figures - P.B.K.)

Complications of the erosion cycle

Great Plains of Kansas and Nebraska

The Great Plains[74] of Kansas and Nebraska are aggradational
features. They are capped by late Tertiary deposits
(Ogallala Formation) - gravels, sands, and clays, cemented
by lime to form 'mortar beds' (= caliche). During the
early geological surveys of the plains, these deposits were
assumed to have been laid down in vast, shallow lakes, but
the plains surface slopes gently eastward from the Rocky
Mountains - from more than a mile above sealevel at the
mountain base to less than a thousand feet above sealevel
at their eastern edge near the Missouri River - which would
require a regional tilt of the former 'lake bottom'. True
conditions were demonstrated by Willard D. Johnson (1901)
in a classic report which showed that the deposits were
built by withering streams, flowing eastward from the Rocky
Mountains and from which they derived their detritus.
These Great Plains are thus piedmont alluvial plains.
     These Great Plains are thus neither the surfaces of
former lake deposits, as first believed, nor are they
plains leveled by erosion, as in Montana (see below).
Erosion is at work on the plains, it is true, but it is
still increasing rather than diminishing the relief. The
main depositional surfaces are still level and untouched,
but the edges of the deposits both toward the mountains and
toward the Missouri, are now being dissected, producing
escarpments known as the Breaks of the Plains.
     Special features of these Great Plains, especially
along the bordering escarpments, are the badlands. These
have been cut in the fine-grained, impervious plains
deposits, in a semi-arid climate which has permitted the
growth of only an impoverished vegetation. The resulting
forms are an intricate network of small, closely-spaced
valleys and ridges, drained by innumerable rills (rather
than streams) - an incident in the early stages of the
cycle of erosion.

Great Plains of Montana

The Great Plains of the basin of the upper Missouri River
in Montana are level-surfaced, but are now being dissected
close to the Missouri by young tributaries. The plains at
first sight give the impression that they are young
aggradational features, as in Kansas and Nebraska, whose
surface was the original surface of deposition. Further
examination shows that the plains are not young, but old,
and that their present surface is not the original one.
     The plains surface is not underlain by a single
stratum, and some of the more resistant, slightly tilted
bedrock units project above the general level as faint
ridges. In places, also, dikes project above the surface
in walls hundreds of feet high, and elsewhere lava-capped
mesas preserve strata higher than those on the plains
surface (Fig 60). Clearly, these lavas once flowed down
low places on an earlier plain which was higher than the
present one. Moreover, at the western edge of the plains,
east of the front of the Rocky Mountains, are the Crazy
Mountains, composed of flat-lying remnants of the higher

Figure 60. A lava-capped mesa in
    the Great Plains of Montana.
    Dotted lines indicate the former
    topographic sag in which the
    lavas flowed (from a blackboard
    sketch).

Figure 61. The Crazy Mountains of southwestern Montana,
    formed of weak, flat-lying latest Cretaceous and
    Paleocene strata, held together by a network of dikes
    (from a blackboard sketch).

units of the original stratified sequence, which have
been preserved because they are held together by a network
of dikes (Fig 61).

Plateaus

Like plains, plateaus are level surfaces, but they stand
higher than plains, hence are more deeply dissected by
valleys or canyons. Plateaus are composed of rocks of
various compositions and structures. In the variety
discussed here, the plateaus are formed of stratified rocks
that are flat-lying or nearly so, and have an initial surface that stands far enough above baselevel so that the
region will have much relief when penetrated by erosion.

Many possible variations in the arrangement of strong
and weak strata can occur in a plateau region, producing
corresponding variations in the surface forms (Fig 62).
However, for purposes of analysis, we can assume an ideal
case, in which there is a succession of hard and soft
strata lying horizontally (Fig 63).

Early stages

Insequent drainage will develop across the horizontal surface of the plateau. The larger streams will begin to
incise the plateau and cut canyons (Fig 63A). These serve
as starting points in the recession of the uplands. In the
early stages the cliff-makers may overhang, if the rock is
sufficiently massive and unjointed (Fig 64A). This is
because the slopes below are so steep that talus from above
will not lie on it and protect it. The lowest cliff-maker
is soon obscured by talus and loses its topographic expression after the initial stages. The upper cliff-
maker, of nearly the same thickness and strength, forms a
cliff until the whole layer is consumed, as there is
nothing above to hide it.

At first the talus[75] of the upper cliff-maker will
reach exactly to the rim of the middle cliff-maker
(Fig 63A, 64A), but as erosion progresses, the talus slope
will retreat from the rim, and a platform will intervene
between the base of the talus slope and the rim of the

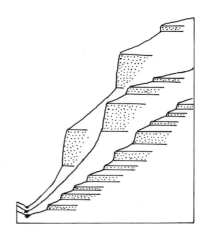

Figure 62. Profiles showing a
few of the many possible
variations in the arrangement
of strong and weak beds which
could occur in a plateau
region (from Davis, 1908,
fig 2b, pl 8).

cliff-maker below (Fig 63C, 64C). This platform is not
developed on the upper bedding plane of the cliff-maker
below, but has a slight gradient away from the base of the
talus and toward the rim of the cliff beneath; this
gradient is adjusted to the size of the materials in transport across it, just as in the pediments of arid regions.
The most striking example of such a platform is the
Esplanade in the Grand Canyon, once thought to represent
an old erosion surface.

In each cliff, there are therefore three elements: the
cliff itself, commonly nearly vertical; the talus slope,
with a constant, rather steep angle, which is the angle of
repose of the large pieces fallen from the cliff; and the
platform, with a rather gentle gradient adjusted to the
size of the materials being transported across it from the
base of the talus to the rim of the next cliff-maker below.
On the cliff, bedrock is bare; on the talus slope it is
thickly mantled with large fragments; on the platform it is
thinly mantled with materials in process of transport.

Later stages

As erosion progresses, the upper cliff retreats most
rapidly, and the platform below it soon becomes very wide
(Fig 65E). The base of the talus slope becomes higher and
higher, as it is adjusted to the headward projection of the
platform gradient. If there is sufficient room for retreat, the platform may eventually intersect the upper
cliff-maker, and extinguish it.

A plateau, trenched by a branchless master river, would
have its cliffs in two long lines about parallel to the
river, and to each other on opposite walls of the canyon.
If the river had tributaries, each will cut back a side
canyon (Fig 66). The tributaries and side canyons will
have angular heads, because the back-cutting of the canyon
head by its streams is faster than the widening of the
canyon by weathering of the walls. The larger the stream,
the sharper the canyon-head angle, and vice versa. As the
canyon head is cut back nearer and nearer to a stream head

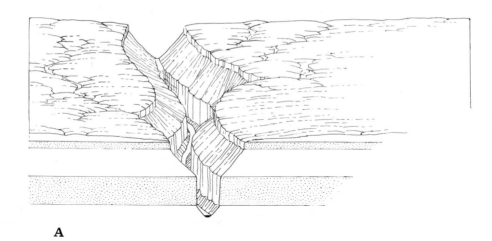

A

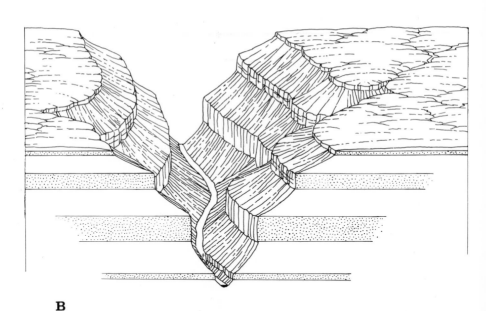

B

Figure 63. An ideal case of the erosion of a plateau region: Four diagrams (A, B, C, and D) showing the progressive development of a canyon by erosion (from Davis, 1908, fig 1, pl 7; fig 2, pl 8; fig 3, pl 9; fig 4, pl 10).

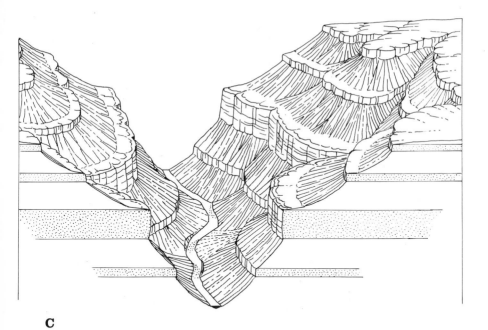

C

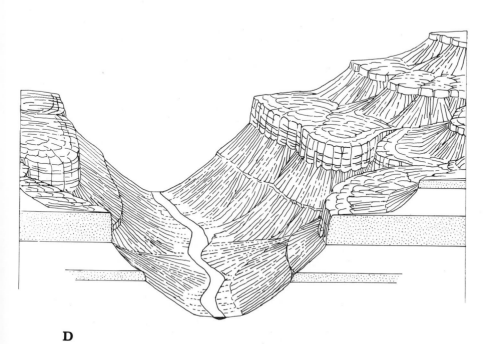

D

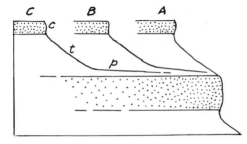

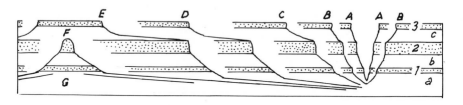

Figure 64. Profiles
showing development of
cliff *c*, talus slope *t*,
and platform *p* (from a
blackboard sketch).

Figure 65. Profiles showing progressive forms developed
during erosion of a plateau region. (from Davis, 1908,
fig 10, pl 8).

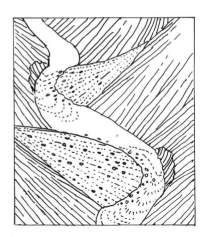

Figure 66. A detail in the
erosion of a plateau
region: Deflection of a
stream in a canyon bottom
by alluvial cones from
side branches, producing
rapids in the stream
(from Davis, 1908, fig 3a,
pl 9).

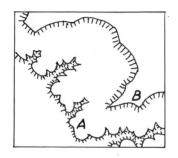

Fig 67. Contrast between
lower scarp *A*, with convex
outward plan, recessed by
stream erosion, and upper
scarp *B*, with concave outward plan, recessed by
weathering (from a black-
board sketch).

it retreats slower and slower; when the stream head is reached, the canyon head is a rounded alcove. With further retreat of a canyon head into the streamless plateau, the alcove is enlarged into a great amphitheater, largely as a result of weathering. Good examples of these features may be seen in the cliff patterns of the Grand Canyon[76] (Fig 67).

In consequence of these systematic changes, the plateau spurs, when first defined by the beginning of side-canyon cutting, have short, heavy, blunt forms between short, narrow canyons. Later on, as the canyons lengthen and widen, the spurs become longer and narrower. Then they are sharpened, sometimes acquiring the form of slender spires between widened alcove- or amphitheater-headed canyons. Later still, as the amphitheaters enlarge, the spine-like spurs are shortened into dull cusps. The outer parts of the spurs are not infrequently cut off from the main body of the plateau as outliers by the growth of tributary branches behind them, then dwindle into outposts, and finally vanish.

In a high plateau, cut by a deep canyon the spurs near the canyon bottom, being the latest developed by the deepening of the main canyon, have short, heavy, blunt forms between narrow side canyons; the spurs at mid-height are further developed and exhibit sharpened or spine-like ends between the round-headed canyons or amphitheaters. At highest level, great amphitheaters are worn back between shortened, dull spur cusps. In a deep-cut sub-mature canyon like the Grand Canyon, the low-lying cliffs still show the pattern that the high-lying cliffs originally had, and the high-lying cliffs show the pattern that the low-lying cliffs will acquire later.

In a young plateau the canyons occupy but a small part of the plateau (Fig 68A). As the cliff-formers retreat, the canyons will widen into flat-bottomed valleys (Fig 68B). These valley lowlands are not the result of undercutting and wandering of streams on their floodplains, but are formed by recession of the cliffs and talus slopes from the stream lines, and the development of gently-sloping platforms between the base of the talus and the streams. Maturity is reached when these lowlands along the streams equal the area of the still-unconsumed uplands.

An example in Utah

In horizontal strata, canyons will often show a progressive sequence of forms downstream, as lower beds of different character are penetrated. A good example is along the Virgin River in Utah, back of the Grand Wash Cliffs (Fig 69).

a) At its headwaters, the Virgin River flows through a thick sequence of cross-bedded desert sandstone of Jurassic age. Its narrow, slit-like canyon in this region has been famous since the days of the earliest explorations; the cross-section was shown on the cover of LeConte's Textbook of Geology.

b) Downstream, weaker Triassic shales (Chinle) are penetrated, which cause the great sandstone cliffs to

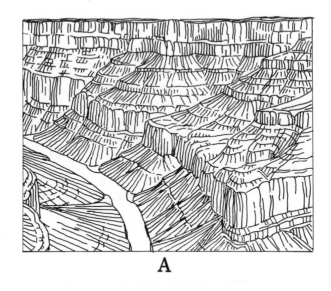

A

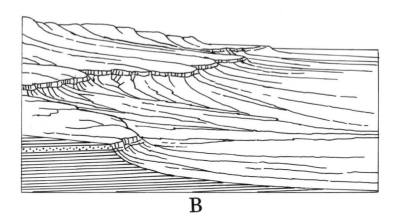

B

Figure 68. Erosional features of plateaus illustrated by
examples in the Grand Canyon region: A) A young,
narrow canyon; Marble Canyon, cut in Paleozoic limestones and sandstones. B) An area of advanced scarp
recession; ledges and slopes cut on Mesozoic formations
north of the Grand Canyon (from Davis, 1912, fig 38).

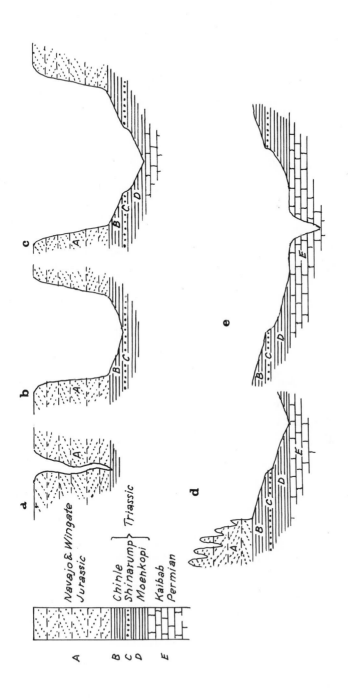

Figure 69. Profiles of valley of Virgin River back of the Grand Wash Cliffs, southwestern Utah, showing progressive sequence of forms displayed in a downstream direction (a through e) (from Davis 1912, fig 40).

71

recede from the river. They are sapped by the undercutting
of the shale, and break off in great sheets, although at
the cliff base there is no talus, because of the rapid dis-
integration of the sandstone. The tributaries have not
reached the weak stratum, and hang above the main valley.

c, d) Lower down, the sandstone is reduced in height
and forms groups of sharp pinnacles on the divides away
from the river. When these pinnacles give way, great landslides result, which in places extend clear to the river's
edge.

e) Near the edge of the Grand Wash Cliffs, the valley
broadens out on the weak Triassic shales, and then an inner
gorge makes its appearance, cut on the underlying Permian
(Kaibab) limestone.

Faulting in plateau area

After displacement, a fault is expressed as a <u>fault scarp</u>,
but erosion soon causes dissection and recession of this
scarp (Fig 70). With great recession, the scarp can no
longer be considered a fault scarp, but is merely an
erosion scarp. In old age, the scarp is obliterated
entirely; there is no topographic expression, but the fault
trace divides areas of contrasting rocks and soils.

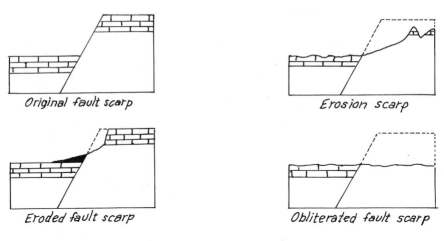

Original fault scarp   *Erosion scarp*

Eroded fault scarp   *Obliterated fault scarp*

Figure 70. Sections showing fault scarps: Original fault
scarp; eroded fault scarp; erosion scarp; obliterated
fault scarp (from a blackboard sketch).

With continued erosion, if the land is high or if it is
in its second cycle, <u>fault-line scarps</u> will develop
(Fig 71). These may be <u>resequent</u>, and face in the same
direction as the original scarp, or they may be <u>obsequent</u>,
and face in a direction opposite to that of the original
scarp. The height of a fault-line scarp is a function of
the erosion of the weak and strong beds in the strati-
graphic sequence, and bears no relation to the amount of
the original fault displacement - contrasting with a fault
scarp, which shows clearly the original throw of the fault.

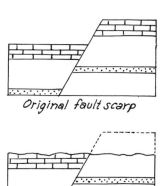

Original fault scarp

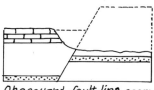

Obsequent fault-line scarp

Obliterated fault scarp

Resequent fault-line scarp

Figure 71. Sections showing fault-line scarps: Original
fault scarp; obliterated fault scarp; obsequent fault-
line scarp; resequent fault-line scarp (from a blackboard sketch).

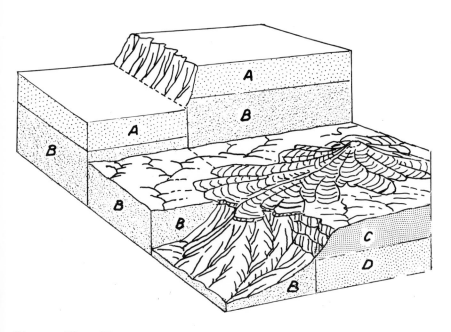

Figure 72. The resequent fault-line scarp of Hurricane
Ledge, north of the Grand Canyon. Proof of its fault-
line nature is afforded by the lava cap, which was
spread over the fault, after its original scarp had
been obliterated by erosion (from Davis, 1912, fig 76).

Complications of the erosion cycle

An example of a fault-line scarp is Hurricane Ledge
(Fig 72), north of the Grand Canyon in the plateaus of
Arizona. This scarp is developed on the faulted bedding
surface of the Permian limestones C, and its fault-line
nature is demonstrated by the fact that it is crossed by
lava flows which preserve, under their resistant capping,
remnants of Triassic red shales B on the downthrown side.
Clearly, the lava flowed over the faulted terrain after
the original fault scarp had been obliterated by erosion,
and had been worn down to a nearly level surface. The
present scarp results from erosion of this surface.

Mountains

Mountains are landforms of great altitude and strong
relief, but the term is vaguely used. Dissected plateaus
(as in West Virginia and southeastern Kentucky) are
commonly called mountains; so are high volcanoes (such as
Shasta and Fuji-yama). The term will be used here for
forms of more or less disordered structure, uplifted to
considerable or great height, and dissected to varied
relief. Mature dissection greatly increases the variety
of relief over that due to upheaval, which ordinarily produces massive forms. Late mature degradation reduces
lofty mountains to peneplains, or plains.

Many mountains are composed of disordered structures
produced by compressional folding and overthrusting. Such
compression would cause an upheaval of the compressed
region. Until the latter part of the 19th century, most
of the great mountain ranges of the earth were thought to
owe their height to such compressional crushing. Later
investigation has shown that all the great ranges of the
earth, so far as studied, are no longer in the cycle of
erosion introduced by their compressional upheaval, but in
a second, or nth cycle produced by broad upheaval,
following the reduction of the first-formed mountains to
subdued or low relief.

Block mountains

Much simpler than those mountains containing structures
produced by compressional forces are block mountains,
generally resulting from crustal tension, in which blocks
of the crust are broken by faults which split the terrain
into blocks of variable width, but generally wide enough
to divide the region into discrete and persistent units.
Because of the faulting and the tilting of the fault
blocks, some project as mountains and others form inter-

75

Complications of the erosion cycle

montane depressions. Unlike other mountains, block mountains (in their early stages at least) are <u>shaped directly</u>
<u>by the structures themselves</u>. However, after the time
of block-faulting, erosional forces more or less modify the
original tectonic forms, as explained below.

Young block mountains of southern Oregon

Some of the simplest and youngest block mountains are in
southern Oregon, described by I.C. Russell (1884) (Fig 73).
Their pre-faulting structure was a crystalline foundation
of low relief, on which a great body of lava had been
spread, producing a lava plain. The faulting broke the
foundation, the lava body, and the lava plain into blocks,
which were jostled up and down like blocks of ice. The
lava plain thus forms the backslope of each mountain range,
the opposite faulted side of which is a cliffed scarp. In
many ranges, faulting and uplift have been so recent that
landslides have broken off and slid down the scarps. The
region is drained by consequent streams, which have performed little dissection since the faulting.

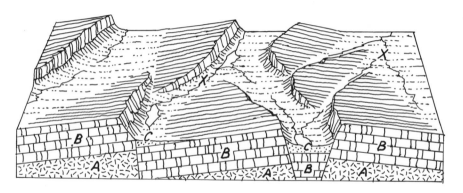

Figure 73. Simple block mountains of the sort displayed in
    southern Oregon. A) Crystalline basement. B) Lavas,
    little deformed before faulting. C) Basin fill.
    X) Consequent stream (from Davis, 1898, fig 101).

    Steens Mountain in this region shows many of these
features, although it is more dissected than the choicest
examples cited by Russell.
    The consequent drainage in an area of young block
mountains will form a simple pattern (Fig 74). <u>Definite</u>
<u>consequent streams</u> will flow along the traces of the
faults, between the backslope of one mountain and the fault
scarp of the next. <u>Indefinite consequent streams</u> will flow
down the slopes of the fault scarps and of the back slopes.
These are consequent because they can flow only in one
direction - down the structural surfaces. They are indefinite because they can take their positions anywhere on
these surfaces.

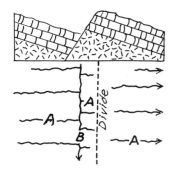

Figure 74.   Types of consequent
drainage resulting from
block-faulting: A) Indefinite
consequents.   B) Definite
consequents (from a blackboard sketch).

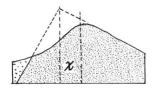

Figure 75.   An obsequent
addition x to a consequent
stream on a block mountain
(from a blackboard sketch).

After erosion begins, the steeply sloping fault-face
consequents will lengthen their drainage headward at the
expense of the backslope consequents, which have much
gentler gradients.   This added drainage constitutes an
obsequent extension of the fault-face consequent streams
(Fig 75).   In the meantime, the depressed parts of the
blocks are being buried under detritus derived from the
upraised blocks.

Old block mountains of the Basin Ranges

A region long-ago deformed and worn down to low relief may
be broken and displaced by block-faulting.   The mountains
thus produced differ from those just discussed in that
their interiors consist, not of rocks of simple structure,
but of complex structure.   The pre-faulting form of the
surface is also likely to be complex, as it may have
reached any age in the cycle of erosion.   In such a region,
three items must be considered: 1) The form of the region
before the faulting.   2) The structural changes accomplished
by the faulting.   3) The changes accomplished by erosion
during and since the faulting.   Item (1) can be further
subdivided into: the structure of the region; the processes
that have operated to modify it; and the stage to which
these processes had been carried at the time when the
faulting began.

The Basin Ranges, lying mainly in the Great Basin of
Nevada and adjacent states, between the Sierra Nevada of
California and the Wasatch Mountains of Utah, illustrate
old mountain fault blocks.

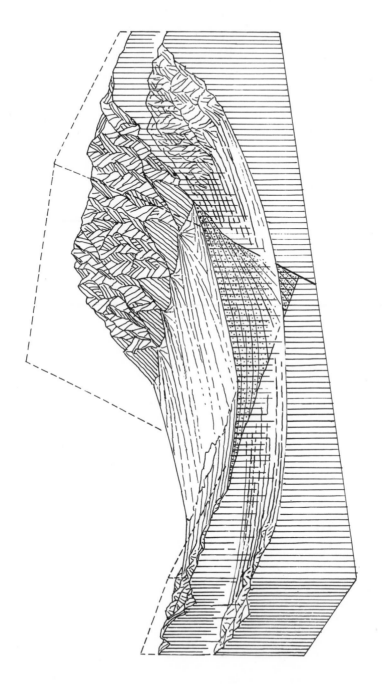

Figure 76. Diagram of block mountains and intervening bolson in the Basin Ranges. Heavy dashed lines show the original form of the tilted blocks. Heavy lines show the region after vigorous dissection of the mountains and filling of the bolson, where a central playa receives interior drainage. Light lines show the region after more prolonged erosion; exterior drainage has been established, much of the bolson fill has been removed, and most of the lowland surface has become a rock floor, or pediment (from Davis 1898, fig 201).

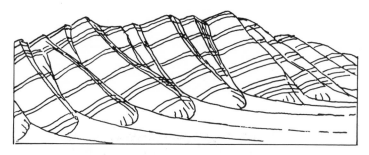

Figure 77. A faulted, even baseline of a mountain front truncates the internal structure of the range at a low angle; Marble Mountains north of Cadiz; Mojave Desert region of southeastern California (from Davis, 1933, fig 7).

They are isolated, nearly parallel mountains with dominant north-south trend, elongated in the same direction, which are separated by plains or bolsons underlain by subaerial deposits derived from the waste of the mountains (Fig 76). The rocks of the mountains are generally much deformed by events prior to the block-faulting. Their present topographic forms must have resulted from this faulting, although the faults themselves are only rarely visible, and the evidence for their existence is largely in the surface forms. Because of the faulting, the mountains end on an even baseline, generally oblique to the strike of the rocks - strong and weak beds alike end abruptly at the edges of the mountains (Fig 77)[77].

The climate of the Basin Ranges is arid and the landforms have been shaped according to the principles of arid erosion. Little of the Great Basin has through-flowing drainage, although south of it in a similar terrain a few rivers drain to the sea - the Colorado, the Gila, the Rio Grande, and their tributaries.

The first comprehensive geological study of the Great Basin was by the Geological Exploration of the Fortieth Parallel (King Survey), which ably demonstrated the stratigraphy and structure of the rocks within the ranges. Clarence King, director of this survey, interpreted the present landforms as resulting from normal processes of erosion on the folds seen within the ranges. Not long afterwards, however, G.K. Gilbert (1875) suggested that the ranges are fault blocks whose forms are unrelated to the folding. At about the same time, J.W. Powell (1875) postulated that the region deformed by the folds was worn down to a surface of low relief before the formation of the fault blocks. Much later, G.O. Louderback (1904) showed that in many of the ranges this surface was overspread by sheets of lava before the block-faulting.

The following shorthand terms are proposed for the various features (Fig 78): A) The King formations. B) The King folds. C) The Powell surface. D) The Louderbacks; and E) The Gilbert fault blocks.

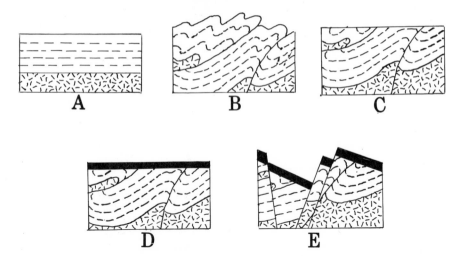

Figure 78. Sections showing the Davis shorthand terms for the evolution of the Basin Ranges: A) The King formations. B) The King folds. C) The Powell surface. D) The Louderbacks. E) The Gilbert fault blocks (from a blackboard sketch).

Nature of the backslope

If an old mountain region were of low relief at the time of the faulting, its streams must have been adjusted to its structure, and must have had directions unrelated to later fault trends. The slant of the backslope would rarely suffice to extinguish such streams and replace them with new consequents, but the slope would greatly affect their competence to erode. For streams which have thus persisted on the backslope of mountain blocks during their tilting, the name persequent[78] is suggested.

Tilting would affect the persequent streams variously, according to the relation between their pre-tilting courses and the direction of the tilting: a) If their courses were in the same direction as the tilting, they will be accelerated, and will entrench themselves in canyons. b) If their courses were in a direction opposite from the tilting, they will be retarded, and they will aggrade their courses to form meadows. c) If they flowed parallel to the strike of the tilting, they will be little affected.

Examples of all three types of persequent streams can be observed in the Inyo and Panamint Ranges of southeastern California.

Such persequent streams are merely temporary features. The accelerated and actively downcutting streams (a) will acquire new drainage at the expense of the retarded streams (b). After submature or mature dissection of the old-mountain fault block, its backslope valleys will be best developed in a downslope direction - except where contrasting strong and weak beds compel them to follow subsequent courses along the strike of the rocks.

Significant evidence for the nearly level nature of the pre-faulting landscape was discovered by G.D. Louderback in and near the Humboldt Range in northwestern Nevada. Here, the older rocks on the backslopes are widely sheeted over by relatively young lavas, tilted and displaced by the block-faulting. These tracts of lava extend over scores of square miles, indicating that the surface over which they spread was nearly flat. Nevertheless, a few places in the region are seen never to have been covered by lava, and may have been hilly to sub-mountainous at the time of the lava eruptions. For the lava covers on the backslopes of the Basin Ranges, the term <u>louderbacks</u>, is proposed in honor of their discoverer[79].

Louderbacks occur widely in the Basin Ranges. Examples include lava patches in the Argus Range east of the southern end of the Sierra Nevada (Fig 79); lava caps on the Piute Mountains northwest of Needles in southeastern

A

B

Figure 79. The Argus Range, east of the southern end of the Sierra Nevada, southeastern California. A) The western, or scarp face, showing remnants of lava caps (louderbacks) standing at different altitudes, indicating step faulting along the range front. B) The eastern, or backslope, showing more extensive remnants of the lava caps, sloping toward observer (from Davis, 1930b, figs 3 and 7).

Figure 80. The eastern, or scarp face of the Piute
Mountains, northwest of Needles, southeastern
California, showing lava cap, or louderback (from
Davis, 1933, fig 8).

A

B

Figure 81. The Galiuro Mountains northeast of Tucson
Arizona, a block mountain with a thick lava cap
(Miocene). A) View of the scarp at about mid-length,
showing lava ledges on scarp and younger basin fill
of San Pedro Valley (Pliocene) in foreground.
B) Sombrero Butte, an outlier of the lava cap in
front of the escarpment, with basement rocks (Precambrian) below it in foreground (from Davis and
Brooks, 1930, figs 3 and 7).

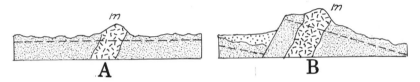

Figure 82. Theoretical sections showing: A) A surface of considerable relief with a monadnock, *m*, cut on the rocks of a pluton. B) The disruption of the surface by block-faulting (from a blackboard sketch).

California (Fig 80), and a thick lava cap on the crest of the Galiuro Mountains northeast of Tucson, Arizona (Fig 81).

Before the faulting of the Basin Ranges, some areas may have been rugged, and some areas may have had rugged relief, with higher monadnocks (Fig 82). However justifiable this deduction may be, evidence for it is scanty because of the considerable later erosion of the ranges - except for indications of pre-faulting relief near the Humboldt Range, as noted above.

The scarps

The faults of the Basin Ranges follow simple courses, yet they are seldom rectilinear. Many have a bight and cusp pattern, with the concavity of the bights toward the downthrown side (Fig 83). It may be that the cusps are localized in more resistant knots of the heterogeneous mountain blocks, yet most of them seem to have little relation to the internal structures of the ranges. The dips of the faults are believed to be steepest in the cusps (perhaps as much as 70°), and least in the bights (perhaps as little as 50°)[80].

After initial displacement, the fault face forms a little-dissected wall, not modified during the strong and rapid faulting. Some of the most imposing of these rock walls are along the eastern side of south-central Death Valley[81]. Later, the wall is notched by valleys so that it rises in triangular facets between the valleys, somewhat worn back from the initial fault face (Fig 84). Alluvial fans are prograded from the valley mouths over the adjacent depressed area.

The streams which flow down the scarps are all consequent, but because of the steepness of their gradients they gain obsequent extensions by rapid headward erosion, at the expense of the streams on the backslope. Where the uplifted block is composed of rocks of contrasting resistance that had been tilted prior to the faulting, subsequent streams develop easily. In places along the front of the Wasatch Mountains south of Salt Lake City are landslides - indicators of very recent faulting.

Renewed faulting

Nearly all the Basin Range faulting occurred at multiple times, rather than during a single time of uplift. This

Figure 83. Plan and section, showing bight and cusp pattern of Basin Range faults (from Davis, 1930b, fig 5).

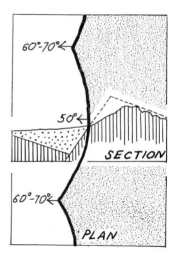

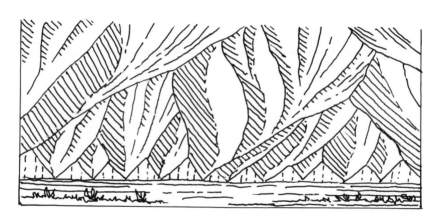

Figure 84. Triangular facets at the foot of the fault scarp of the southern Wasatch Mountains, southeast of Provo, Utah (from Davis, 1903).

progressive uplift brings about many characteristic land-forms (Figs 85, 86, and 87). Faulting may be renewed and tilting increased at any stage in the erosion of the ranges. In ranges in an arid climate, such as the Basin Ranges, with maturely dissected fault faces and backslopes, large fans spread out in the troughs between the ranges. When faulting is renewed, the mountain block rises and the trough block sinks, and each block is given a greater tilt than before; each range gains a new fault face below its maturely dissected fault face of an earlier uplift. The raised valleys of the fault front are incised, and the depressed fans are buried under the new fan deposits. The previously maturely carved valleys of the fault face will now stand with hanging mouths above the trough floor below

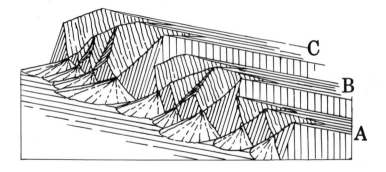

Figure 85. Topographic forms produced during a progressive
    uplift of a fault scarp, not interrupted by a notable
    period of stillstand (from Davis, 1903).

Figure 86. The effect of renewed uplift on a fault scarp,
    after a notable period of stillstand. In A, the
    scarp is fan-bayed; in B, after faulting the scarp
    is fan-based, with drainage off the scarp through
    'hour-glass valleys'. Compare with Fig 92 (from
    Davis, 1903).

them; the wet-weather streams from the hanging valleys will
cut deep and narrow clefts ('hour-glass valleys'). Renewed
faulting is well displayed in Death Valley as well as in
Panamint Valley, next to the west, in southeastern
California.
    Detritus will be washed down into the depressed
troughs, especially from the steepened backslope, so that
the detrital surface of the trough will slant from the
backslope of one block to the new fault face of the next.
Fans from the backslope will soon cover the depressed fans
below the new fault face. By the tilting and the detrital
filling, the central playa of salt pan will be shifted
toward the new fault scarp and may even lie against its
base for a time, and the small fans at the base of the
scarp will not be able to push it away (Fig 87C). A playa
in this position indicates recent movement, and also
suggests that both mountain and basin blocks shared the

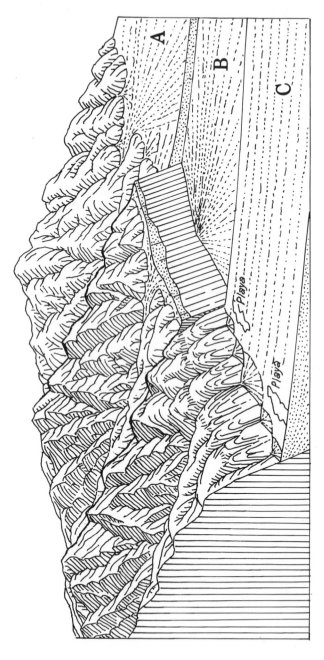

Figure 87. The effects of renewed uplift on a fault scarp: A) A maturely eroded fault block, with a large fan outwashed from its chief valley into the adjoining fault. B) The potential form of the same, after further upfaulting of the mountain block and depression of the trough ; a slanting fault scarp separates the two parts. C) The same, after moderate erosion and deposition consequent on the renewed faulting (from Davis, 1930c, fig 4).

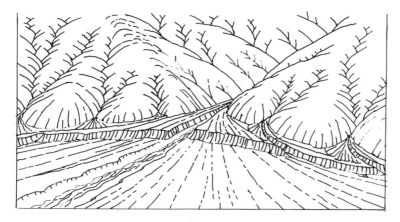

Figure 88.   A faulted alluvial fan at the mouth of a
    valley, in a maturely dissected fault-block mountain.
    In consequence of recent renewal of faulting, the fan
    is traversed by an 'eye-brow scarp' (from Davis, 1930c,
    fig 5).

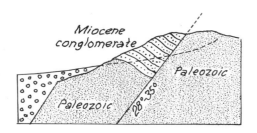

Figure 89.   An accessory fault block which has raised a
    fragment of an earlier basin fill; east side of Cache
    Valley, Utah (from a blackboard sketch).

movement, rather than that the mountain block rose and the
basin block remained stationary.
    The new fault plane is often steeper than the earlier
one, and thus its surface emergence is in front of the
earlier fault.   When faulting is mild, this results in
scarps in the bolson deposits (Fig 88); when faulting is
severe, alluvial deposits are carried up to form high
alluvial benches, that are eventually eroded away (Fig 89).
Scarps in alluvial fans are prominent below Wildrose Canyon
in Panamint Valley, California, and also along the front of
the Wasatch Mountains, Utah.

Effects of prolonged erosion

If prolonged quiesence occurs after the faulting, the fault
scarp and the mountain block will pass through a regular
sequence of erosional forms (Fig 90)[82].

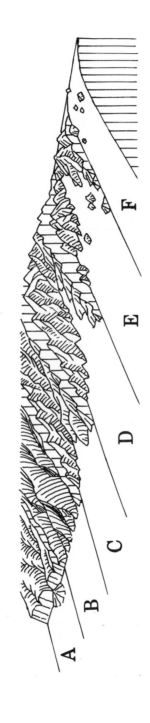

Figure 90.   Diagrams showing stages in reduction of a mountain mass of Basin Range type. Stages are as follows: Fan-free (now shown).   A) Fan-based.   B) Fan-dented. C) Fan-bayed.   D) Fan-frayed.   E) Fan-wrapped.   F) Pan-fan (from Davis, 1925, fig 2).

A) At first, fans will occur at intervals along the scarp, at the points of debouchure of consequent valleys, so that the scarp is fan-based (Fig 90A). B) With further erosion, the fans rise high enough to indent the mountain baseline (Fig 90B). C) Then, the spur ends lose their facets and are rounded off, the valley mouths are enlarged and the fans embay the mountain base; however, the spur ends may still be nearly enough in line to suggest that their terminations were originally along a fault (Fig 90C). D) Later, the fans encroach on the mountain mass with so many branch fanheads that the mountain mass is fan-frayed; now, the detritus on the frayed fanheads is a mere veneer over a baseleveled rock floor, or pediment (Fig 90D). E) Then, the interfan spurs are more consumed and adjoining fans become confluent, leaving isolated inselberg or monticules between, so that the mountains are fan-wrapped (Fig 90E). F) Finally, the whole mountain mass is reduced to a pan-fan - a dome-like residual, everywhere with a graded slope on a thinly waste-covered rock surface (Fig 90F). Examples of the end-stage occur in the Mojave Desert region of southeastern California.

Intermontane depressions

The depressions in the Basin Range province, the original structural low places, are known as bolsons, basins, or valleys. They are filled with all the waste derived from the erosion of the mountains, minus that part which has been exported by exterior stream drainage, or by the wind. Although erosion in the uplifted mountains caused them to become more rugged, deposition in the depressions spread a cover of detritus over the originally rugged structural floor, making it more level.

The borders of the basins are marked by alluvial fans extending from the valley mouths. They may grow to radii of 5 to 7 miles and to a height of 1000 to 1500 feet at their apices; commonly they merge into a piedmont alluvial slope, of bajada.

In the central part of the basin is the playa, or ephemeral lake or salt pan, which has shifted this way and that by the advance of alluvial fans from various directions, or by tilting of the whole depressed block. Such a shift in most places is toward the fault-scarp side of the basin.

Age of faulting

Some of the Basin and Range faulting must have extended well back into Tertiary time, but there has also been very recent faulting, such as that which produces scarps in the alluvial fans, and that which has shifted the playas.

The age of the last great faulting in the Salt Lake City area of Utah can be calculated as follows: On the western backslope of the Oquirrh Range is tilted, dissected, and furrowed outwash (Fig 91). The boulders which were washed out during the original erosional period that followed the block-faulting are all rotten, so that they crumble at a blow of the hammer. Along this furrowed slope

Figure 91. Alluvial slopes
on the west side of the
Oquirrh Range, Utah, cut
into by the Bonneville
beach (from a blackboard
sketch).

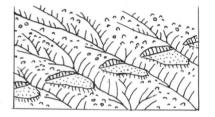

is cut the shoreline of Pleistocene Lake Bonneville, with
wave-cut beaches and gravel bars across the valleys. The
Bonneville gravels are fresh and unweathered.

If we assume B years since Bonneville time, then the
formation of the gravel slope must have been at least 50B
years ago, and the time taken to excavate the valleys in
the mountains must have been 50 times 50B. The time necessary to erode the whole range is probably 100 times 50
times 50B, which evidently represents the length of the
erosion cycle in the region[83].

Objections to theory of block-faulting

Gilbert's deduction of the fault-block origin of the Basin
Ranges was based largely on the surface morphology, a novel
method at the time which aroused much skepticism. Various
geologists have objected to the fault-block interpretation
of Basin Range structure and geomorphology - notably Spurr
(1901) and Keyes (1909, 1912). Keyes published many papers
advancing the hypothesis that the basins are the result of
deflation. Other geologists have rejected parts of the
concept, and undoubtedly various modifications of the
general concept will be required from one part of the
region to another, to satisfy local conditions.

The main difficulty in interpreting the Basin Ranges is
that they are fault blocks tilted into mountains, whose
depressed areas are covered to depths of thousands of feet
by detritus. The bedrock structure of the downthrown blocks
is thus never observable, and only rarely can one even see
the surface of the fault along the mountain border. However, many faults occur within the ranges which are of
diverse ages. Some are older than the block-faulting and
the formation of the Powell surface. A few near the mountain border were formed at the same time as the border
fault, and are accessory to it; from these may be gained an
idea of the nature of the main faulting.

In studying the Basin Ranges it must be realized that
they have been eroded by arid processes. Thus, although
faulting can produce steep mountain fronts during the early
stages of the erosion cycle, other steep mountain fronts
may form later in the cycle by erosion alone, behind eroded
pediments, or rock floors.

Best examples of mountains whose forms have been shaped
primarily by arid erosion are in the Papago country of
southwestern Arizona, described by Kirk Bryan (1922).
Here, the original structural shaping of the mountains -
presumably by block-faulting - took place much longer ago
than in most of the ranges of the Great Basin farther

north, so that they have been frayed and embayed, and their earlier block-like forms have been lost.

Folded mountains

First cycle in folded mountains

To study folded mountains, we can begin with an ideal case (Fig 92): A thick sequence of rocks, composed of alternating strong and weak beds, is deformed by folding. The uppermost stratum is strong - if there were a weak upper stratum, it would have been rapidly stripped off so that the assumption of a strong bed at the top will be true in any event. Folding of the rocks progresses so rapidly that all drainage is consequent on the folded surface. The folds plunge in various directions, so that their axial surfaces rise and fall gently along their length.

The axes of the synclines will be followed by longitudinal synclinal consequent streams, flowing down the plunges toward the lowest sags in the troughs (Fig 93). These will remove from their sides many indefinite lateral consequents, which flow from the consequent anticlinal divides down the consequent anticlinal side slopes. If the folding is rapid, consequent lakes will form in the lowest places in the synclines; if the folding is less rapid, these lowest places will be filled by detritus. The lakes will overflow across the lowest anticlinal sags in the plunging folds, producing transverse, or cross-axial consequent streams.

As the lateral consequents have a strong fall, they will erode rapidly, building fans at their bases which will push aside the axial synclinal consequents. On the slopes, they will soon breach the upper strong stratum, and reach the weak strata beneath, after which the strong upper stratum will be undercut. Erosion may be sufficient to allow some of the lateral consequents to cut headward across the anticlinal axis, so that they will acquire obsequent extensions. Breaching of the upper strong stratum will split the original axial divide into two subsequent monoclinal divides, one on each flank. Axial subsequent streams will now develop along the axis on the weak beds, and will receive obsequent wash from the inner slopes of the two subsequent ridges. Obsequent erosion will cause down-dip retreat of the two flanking monoclinal ridges, and the lateral consequents flowing on the outer slopes will be thereby shortened.

Let us now return to the axial synclinal consequents. These will deepen their valleys slowly, as they have a gentle fall, are much burdened by detritus from the laterals, and are floored by the uppermost strong stratum. Those of their tributaries which lie nearest the cross-axial consequent at the sag in the plunge of the folds will have the greatest advantage, and will erode more rapidly in the anticlinal areas than those farther up the plunges of the folds (Fig 93).

In time, the various branches in the original strong outer surface of the anticlinal ridge will become confluent,

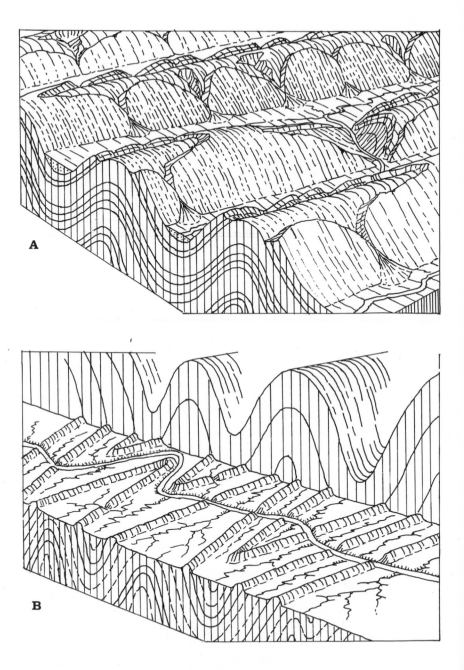

Figure 92. Typical folded mountains: A) Young folded mountains; the Juras of Switzerland. B) Old folded mountains; the Appalachians of Pennsylvania. The view in A is on a larger scale than that in B (from Davis, 1898, figs 105, 118).

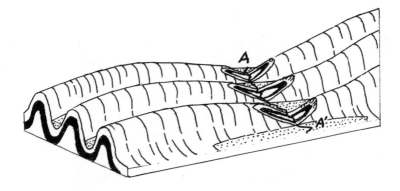

Figure 93. Plunging folds, with a cross-axial consequent
stream (A - A') developed at the lowest sag in the
plunging folds (from a blackboard sketch).

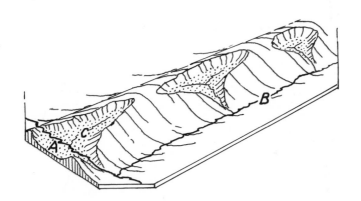

Figure 94. An anticline after moderate dissection, showing
relation of cross-axial consequent, A, to axial synclinal subsequent B, and to axial anticlinal subsequent
C (from a blackboard sketch).

and the ridge will be split all along its crest. Some of
the breaching streams will drain to one side of the anticline, some to the other. Subsequent streams in the two
adjoining breaches will head against each other and will
compete for drainage. The advantage will be held by those
subsequents which drain into those lateral consequents
nearest to the cross-axial consequents (Fig 94).

About this time a second strong stratum beneath the
weak beds will be discovered by deepening of the lateral
consequents and their tributaries (stage 3, Fig 95). This
will be laid bare first near the lateral consequent itself,
and then farther and farther along the anticlinal axis in

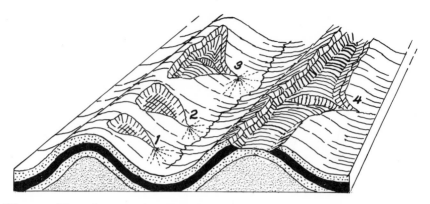

Figure 95. Stages in the breaching of an anticline
(1-2-3-4). A weak bed (solid) lies between two strong
beds (stippled)(from a blackboard sketch).

either direction. This strong core will rise as a resequent divide, shedding lateral consequent streams on
either side (Fig 95, stage 4; Fig 96). The original axial
subsequents will then be split into two subsequents on
either side of the resequent divide, each of which follows
a belt on the weak stratum. As erosion progresses, these
subsequents will sidle down the dip of their guiding weak
stratum. Because of all these adjustments, the consequent
drainage which originally developed on the surface of the
folded terrain will be greatly reduced.

The original axial synclinal consequents, already
cutting downward with difficulty, will thus lose most of
their tributary drainage and will become even less competent to erode the strong stratum over which they flow.
If the region stands well above baselevel, the subsequents
will degrade below the level of the diminished axial consequents, so that they will lose the last of their water.
What was originally a synclinal valley will then become a
synclinal ridge; the anticlinal ridges that originally projected on either side will be worn down into valleys below
its level (Fig 97). This will occur first near the highest
places on the plunges of the folds. The strong cores of
the anticlines will be most exposed where the anticlinal
axes rise the highest - that is, along the residual anticlinal ridges.

There is an erroneous notion that, in much eroded
folded mountains, the ridges are dominantly along the axes
of synclines. Actually, once youthful stages of erosion
have passed, axial synclinal ridges and axial anticlinal
valleys may develop at any stage, depending on the nature
of the local sequence and structure[84]. Moreover, in such
an eroded terrane the ridges can have varied structure; in
Pennsylvania the kinds of ridges in order of abundance
are: first, monoclinal; second, anticlinal; and third,
synclinal.

If no cross-axial subsequent streams develop, the
subsequent streams will have no advantage. This might

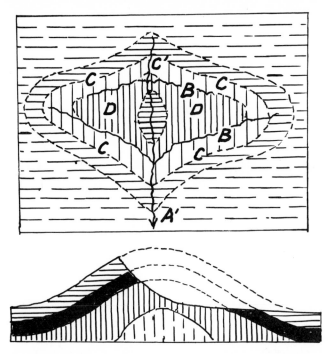

Figure 96. Plan and section of a fold breached by a lateral consequent stream A. B) Subsequent streams. C) Obsequent drainage areas. C') An obsequent extension of the consequent. D) Resequent axial divide (from a blackboard sketch).

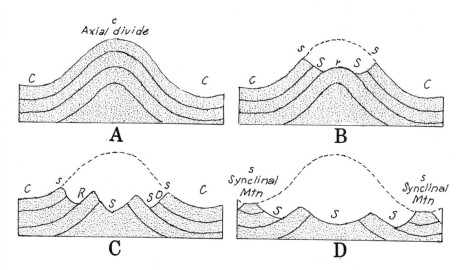

Figure 97. Stages in the reduction of a folded region by erosion, showing changing forms of the ridges and valleys, and their terminology. c) Consequent ridge. s) Subsequent ridge. r) Resequent ridge. C) Consequent valley. S) Subsequent valley. R) Resequent valley (from a blackboard sketch).

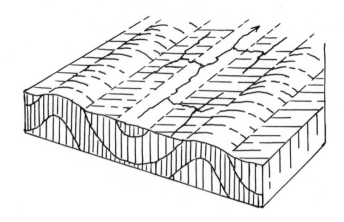

Figure 98. A folded region at the end of the first cycle, showing drainage pattern if there had been no development of cross-axial consequents (from a blackboard sketch).

occur if the folds have little or no plunge. If for this reason (or any other) development of subsequent streams is retarded, the master streams until the end of the first cycle will remain the axial consequent streams, and these will be fed by lateral consequent streams, and by small subsequent streams (Fig 98).

Second cycle in folded mountains

The features and events so far reviewed all belong to the first cycle of the erosion of a folded mountain region. In old age of such a cycle there will be only pale ridges on the strong beds, and if sufficient impetus has been given by the cross-axial consequent streams, the tributary drainage will be largely subsequent.

With the beginning of a new cycle of erosion, adjustments of drainage to strong and weak beds will be greatly accelerated, as in coastal plains; the strong-rock ridges and weak-rock valleys will then become increasingly prominent. At this stage, the cross-axial consequents will find it difficult to cut downward, as in their synclinal crossings they will be eroding the strong upper stratum still preserved along the synclinal axes. Their courses will thus be obstructed by a succession of these strong-rock belts (Fig 99). Nevertheless, these streams have become the dominant drainage lines during the first cycle, and have acquired much volume from tributary consequents and subsequents; generally they will thus succeed in surviving the obstacles of these crossings.

If the folds plunge steeply, the weak-rock belts will have a zig-zag pattern, crossing the axes of the folds. Cross-axial subsequents can develop in these weak-rock belts. Their tributaries will drain the adjacent weak-rock areas - down the plunge along the anticlinal noses, and up

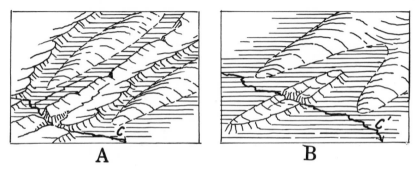

A B

Figure 99. The vicissitudes of a cross-axial consequent,
C-C', during the second cycle in a folded mountain
region, when it must reduce a synclinal barrier athwart
its course (from a blackboard sketch).

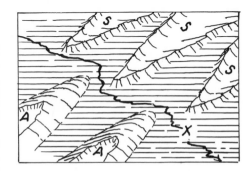

Figure 100. A cross-axial
subsequent stream, X,
developed on weak strata
in plunging anticlines,
A, and synclines S (from
a blackboard sketch).

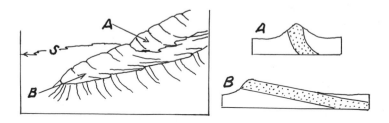

Figure 101. A subsequent stream crossing a strong-rock
ridge on the flank of a syncline. The sections
demonstrate that there is less width of strong rock to
cut through on the flank, A, than along the axis, B
(from a blackboard sketch).

the plunge in the synclinal troughs (Fig 100). They may
also be able to capture drainage beyond the bordering
mountain ridges, by tributaries entering from the sides
rather than the axes of the folds (Fig 101). Here, there
is less strong rock to cut through than along the axes; the
dip is steeper, so that the outcrop belts are narrower.

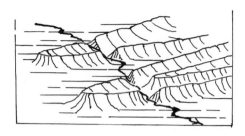

Figure 102. The Susquehanna River above Harrisburg,
    Pennsylvania, crossing the ends of zig-zag ridges,
    apparently as a result of superimposition (from a
    blackboard sketch).

Each cycle will give the subsequent streams greater
advantage, so that more and more of the drainage will flow
along the weak-rock belts. In a region of folded mountains
in the plural cycle, any stream pattern which deviates
from this is very likely to be superimposed. Superimposition of drainage is possible only by cutting through a
cover of superincumbent sediments of different structure,
although some geomorphologists have believed that it could
be produced also by renewed downcutting of streams that had
been wandering at will on a peneplain cut on the folded
rocks themselves.

Superimposition of streams through a sedimentary cover
would involve the following history: A) Peneplanation of a
terrain of rocks of folded or otherwise complex structure.
B) Accumulation of sediments on the peneplained surface,
either continental or marine, the latter if the surface
had been submerged beneath the sea. C) Emergence of the
area as a coastal plain, drained by consequent and insequent streams. D) Renewed downcutting, resulting in incision of the streams and stripping of the sedimentary
cover to reveal again the complex terrain beneath. On
these complex rocks, the incising streams are superimposed,
without regard for their structural pattern.

An apparent example of superimposition is the
Susquehanna River immediately upstream from and north of
Harrisburg, Pennsylvania (Fig 102). Here the river cuts
across the ends of several zig-zag monoclinal ridges.
These zig-zags were probably once covered by the feather
edge of Cretaceous or Tertiary deposits of the Atlantic
Coastal Plain, on which the Susquehanna took its original
consequent course[85].

Erosion features of folded mountains

Appalachian region

The Appalachian region is one of the best examples of old
folded mountains, now in a plural cycle of erosion (Fig 92).
The Appalachians have had a complicated history, and have

been through many erosion cycles[86]. The first consequents
and subsequents probably drained northwestward, away from
the strongest and highest folds of the belt. Later,
drainage was reversed to the modern courses, dominantly
toward the Atlantic - probably by an uplift of the plateau
country to the northwest and a subsidence of the Piedmont
to the southeast.

The Hudson, Delaware, Susquehanna, and Potomac Rivers
rise in the innermost hard-rock belt (the Allegheny
Plateau), where the Ohio headwaters opposite to them flow
west. However, the axis of uparching of the Appalachian
belt departs a little from the trend of the rock belts, so
that the New (= Kanawha) and several branches of the
Tennessee River, both of them tributaries of the Ohio,
rise in a hard-rock belt well to the southeast of that in
which the above-named northern tributaries rise, and cross
several hard and weak belts on their way northwestward.
Branches of all these rivers have developed subsequent
valleys where they cross weak belts; and the intervening
hard units stand up as narrow, even-crested ridges which
abound from Pennsylvania to Alabama. The broadest of the
subsequent valleys is the Great Valley - a true valley, as
it is the product of erosion under the guidance of rivers,
but it consists of many drainage areas separated by low
divides.

Many erosion cycles in the Appalachians during the
Cenozoic, and perhaps earlier, are indicated by preserved

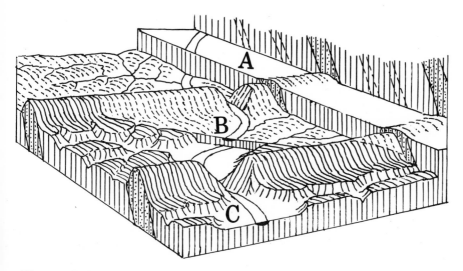

Figure 103. Diagram of a three-cycle Appalachian landscape. A) The smoothly degraded lowland of the first
cycle is shown, broadly uplifted, in an unshaded strip
to the right. B) The inter-ridge, weak-rock lowlands,
worn down in the second cycle between the hard-rock
ridges, are shaded with dotted lines. C) The valleys
eroded in the inter-ridge lowlands after a second
regional uplift are shown unshaded in the foreground
below the full-lined uplands and ridges (from Davis,
1930c, fig 1).

Complications of the erosion cycle

remnants of former level country. Remnants of three main
cycles are preserved (Fig 103), with probable subcycles
between each. The first and oldest forms the <u>summit
peneplain</u> (also called the Schooley), and is preserved on
the tops of the hard-rock ridges. The second and more
extensively preserved forms the <u>valley-floor peneplain</u>
(also called the Harrisburg) and extends across the intervening lower country, where it stands above modern
drainage. The third forms the present drainage level.

<u>Other mountains</u>

Rocky Mountains of Colorado

In the Rocky Mountains, as in the Appalachians, there is
abundant evidence for several cycles of erosion[87]. In the
Front Range of the Rocky Mountains of Colorado, the core
is formed of Precambrian gneiss and granite, and the east
flank descends steeply across sharply upturned Paleozoic
and Mesozoic sediments (Fig 104). The highlands in the
granite core, however, are broad plateaus rather than
mountains. The highlands are rolling and coarse-textured,
with projecting dome-like hills and low mountains (Fig 105).
The east sides of the domes are protected from the prevailing winds so that snow can accumulate, and here glacial
cirques were cut during Pleistocene time. At the margins
of the Front Range, the highland surface is dissected by
youthful canyons of the present erosion cycle.

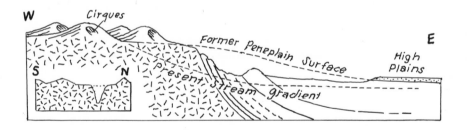

Figure 104. Profile across eastern side of the Front Range
    of Rocky Mountains in Colorado, showing the relation
    of late Tertiary peneplain to present gradients. The
    transverse profile at the lower left shows valley-in-
    valley forms at eastern border of granite area (from a
    blackboard sketch).

    Projection of the old highland surface in the core of
this and other ranges of the Rocky Mountains would carry
it downward to the waste-mantled, high-level surface of the
Great Plains. Generally, the two parts of this old warped
peneplain have been disconnected by later erosion[88], but in
places they are joined by <u>bridges</u>. On the eastern side of
the Laramie Range in Wyoming (the northern extension of the
Front Range), such a peneplain bridge (sometimes called

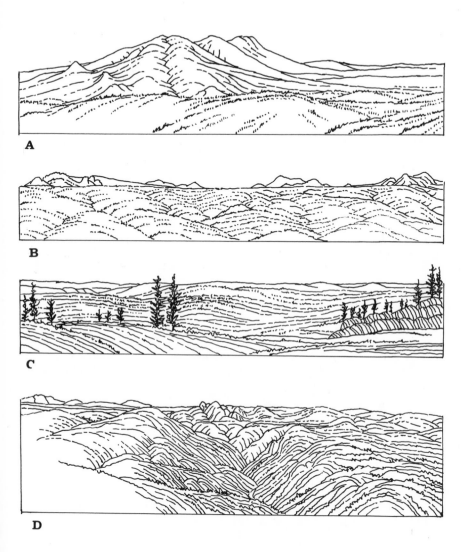

A

B

C

D

Figure 105.   Scenes in the Front Range of the Rocky
     Mountains south of Denver, Colorado. The upper picture
     shows one of the low mountains that project above the
     high upland surface in the granitic core area; the two
     middle pictures show more level parts of the surface;
     the lower picture shows dissection by modern drainage
     (from Davis, 1912, figs A1, A3, C1, and C4).

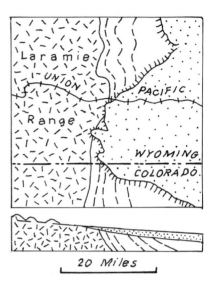

Figure 106. The pene-
plain bridge, or
'gangplank', on the
line of the Union
Pacific Railroad on
the eastern flank of
the Laramie Range,
Wyoming (from a
blackboard sketch).

'the Gangplank') serves as the means by which the Union
Pacific Railroad ascends from the Great Plains to cross the
mountains (Fig 106). In the Black Hills, the uplifted,
mature White River surface (Oligocene) forms high summits
in the core, and in the southeastern hills is connected
with the main body of the White River deposits of the Great
Plains by flat-topped, sediment-covered ridges that bridge
across the Red Valley, the Dakota Hogback, and other
features that are strongly etched elsewhere around the
periphery (see Black Hills folio).

Tian-Shan Mountains

The best example of an ancient, greatly uplifted former
peneplain in a mountain region is the Bural-Bas-Tau, in the
Tian-Shan Mountains of central Asia, where massive granite
has an evenly planed surface, standing at altitudes of
10 000 or 12 000 feet (Fig 107). This surface is obviously
a former peneplain, and at the ends of the range it slopes
down off the sides, thus indicating the configuration of
the uplift which produced the modern range (Davis, 1904).

Coast Ranges of California

The central Coast Ranges of California are formed of a
great thickness of stratified formations, with associated
volcanic beds, which were deformed and much worn down. The
worn down mass was split into northwest-southeast-trending
blocks of unlike length and breadth, and the blocks were
diversely displaced by faults; the upthrown blocks were
dissected and the downthrown blocks were aggraded.

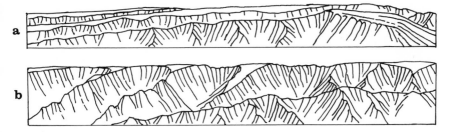

Figure 107. A former peneplain, now preserved only as
    lofty erosion remnants; crest of Bural-Bas-Tau, in the
    Tian-Shan Mountains of central Asia, sketched from a
    great distance (from Davis, 1912, fig 67).

Berkeley lies at the base of an uplifted block, whose
prefaulting surface was probably of subdued form; it is
separated from the non-uplifted or slightly depressed block
on the west by a slanting fault, the scarp of which is now
maturely dissected by many indefinite consequent ravines.
The ranges hereabouts were traversed by a trunk stream
which carried the united Sacramento and San Joaquin Rivers
westward, crossing the uplifted blocks in narrow valleys,
and flowing across the plains surfaces of the intermediate
aggraded troughs. A moderate subsidence caused the submergence of the cross valleys in Golden Gate and Carquinez
Strait, and the submergence of the aggraded trough plains
in the broad San Francisco, San Pablo, and Suisun Bays.
The rivers are now building deltas into the bays.

PART 3　THE MARINE CYCLE

Shorelines

Shore processes

The principal forces at work on a shoreline are <u>waves</u> and
<u>offshore currents</u>[89]．Waves primarily perform erosion, and
do little transporting of the material eroded from the
shoreline; storm waves will carry the material out to
deeper water, but most waves will merely dance the material
outward and inward.　Currents, on the other hand, do little
erosion, but perform much transportation, taking the
material supplied by the waves and carrying it along the
shore.

Storm waves roll in from some other region, and pile
up on the shore in <u>surf</u> (Fig 108A); surf is rarely the
result of local storms and winds, although it may develop
if a local wind blows strongly toward the shore.　As the
storm waves advance toward the shore, the wind behind them
will blow water up their backslopes, increasing their
orbital size and their speed.　The water blown up the backslope will break over the front as <u>combers</u> (Fig 108B).
When the waves move about as fast as the wind itself, the
wind will little affect them and there will be no combing.

Besides storm waves, there are <u>swells</u>, or rounded
waves, some of great strength, but of smaller amplitudes.

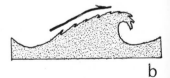

a   b

Figure 108.　a) Surf.　b) Comber (from a blackboard sketch).

104

Swells are a reflection of some far-distant storm, and may
travel thousands of miles from the storm's center.
When the waves reach the shoal water near the shore,
they develop friction with the bottom. The base of the wave
action may be as deep as 20 to 40 fathoms[90]; wave-built
terraces end at about 50 fathoms[91]. Friction on the bottom
strengthens the waves; the waves ahead are crowded by those
behind, the troughs are deepened, and the crests heightened.
When the waves reach the shore they break with a splash as
surf, exerting tremendous energy.

Shorelines of submergence[92]

If we assume that a land of rather rugged relief and a
dendritic drainage pattern be submerged, one of the contour
lines on the former land surface will become the new
shoreline. The sea will drown the lower ends of all the
valleys, dismembering their tributaries, and will advance
around the hills (Fig 109A). As the waves approach such a
shoreline, they will be refracted around the shoaling
headlands, and be bent into the inlets (Fig 110). There
will thus be a great concentration of waves on the headlands, and dissipation of waves in the inlets; for this
reason, inlets make good harbors.
Concentration of the waves on the headlands causes them
to be the most strongly eroded (Fig 109B, C). On a newly
submerged headland, the waves will find surfaces with
smoothly rounded slopes, covered with weathered soil. This
soil will be quickly stripped away from the exposed places
and the fresh rock beneath will be attacked, with differen-
tial erosion of the strong and weak beds. The shore at
the headlands will become a rocky, minutely jagged cliff
(Fig 111). The waves will pull back the detritus as far
from the shore as possible, producing a wave-built terrace
in deeper water . At the foot of the cliff, the waves
will cut out a narrow rock platform.
In early stages, the jagged cliff will be vertical or
even overhang, as wave action then exceeds subaerial
weathering of the slopes above. Later, detritus will
accumulate at the foot of the cliff to form a beach, and
the wave-cut platform of bare rock will be covered
(Fig 112). Wave action will then be slowed to less than
the rate of weathering above, so that now the cliff wall
will be balanced between delivery of material by weathering,
and of removal of material by the waves. This stage is
analogous to maturity in the cycle of valley development,
and the formation of a beach resembles the formation of a
valley floodplain.
The shoreline of southeastern England is in a stage
immediately before maturity; it has ragged coastal cliffs,
abundant coves without beaches, and many caves and chimneys
(Fig 112A, stage 1; Fig 113).
Beaches first appear in the coves, which are either the
lower ends of drowned valleys, or in weak-rock areas in
cliffs that have been excavated by the waves (Fig 112A,
stage 2; Fig 112B; Fig 114). As the rocky promontories are
worn down and destroyed, the coves gradually integrate, and

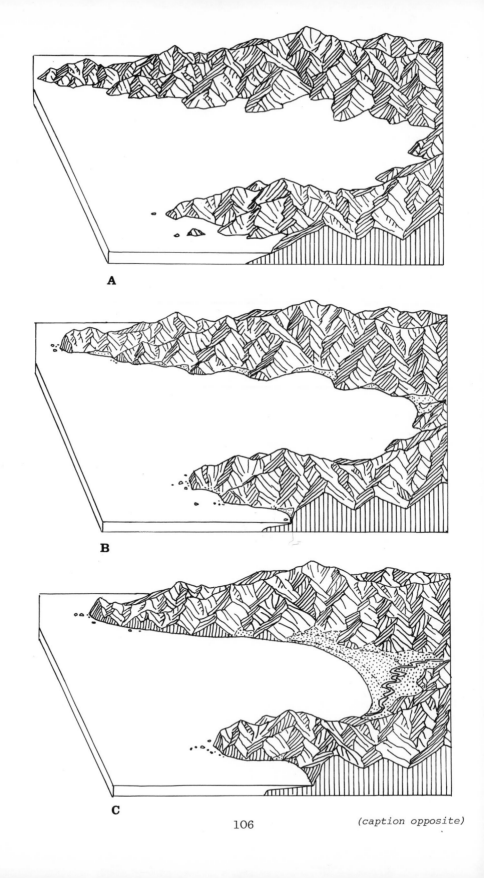

A

B

C

106

(caption opposite)

Figure 110. Refraction of waves on an indented shoreline (from a blackboard sketch).

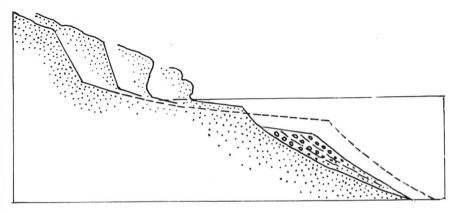

Figure 111. Stages in the development of a sea cliff from initial form, through overhanging, to back-leaning (from a blackboard sketch)[93].

finally the beach is continuous (Fig 112A, stage 3). The offshore currents give the beaches smooth curves, as they cannot turn sharply.

The beach zone will be deepened gradually by the rasping of the waves, using the beach gravels as their weapons, and deepening will be slower at first than back-wearing of the sea cliffs, although the rate of cliff recession will also diminish with time (Fig 115). Thus, gravel and finer material will accumulate faster than normal waves can remove them. They will loiter during quiet periods, but storm waves will sweep them outward.

Figure 109. Three diagrams illustrating the early stages of development of a shoreline of submergence: A) The shoreline shortly after submergence, before much work has been done by waves or streams. B) After minor cliffing of headlands and filling of bay heads by streams. C) After extensive cliffing of headlands and filling of bay heads by streams (from Davis, 1898, figs 228, 231, 232).

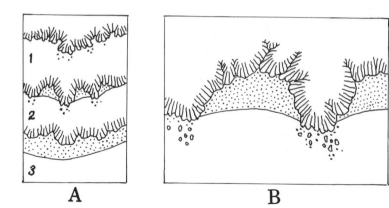

Figure 112.   Stages in the formation of beaches.   In B,
   stage 2 is shown on a larger scale (from a blackboard
   sketch).

The materials thus migrate very slowly into deeper water,
where they will come to rest below wave base[94] (Fig 116).
   While the beaches are growing on the headlands, different processes are at work in the bay heads.   Here the
water is quieter, and the streams that debouche into the
bays will build deltas of land-derived detritus.   Water in
the bays will move in eddies as a result of the offshore
current.   A backset eddy will form in the outer part of
the bay, and farther back there may be a foreset eddy,
moving in the opposite direction[95] (Fig 117).
   Between backset and foreset eddies will be a dead-end
triangle that can be filled with sediments, producing a
cuspate foreland of Gulliver, or a V-shaped embankment of
Gilbert, who observed them along the ancient shorelines of
Lake Bonneville (Fig 118).   Dungeness on the southeast
coast of England is a prominent cuspate foreland, built of
successive gravel ridges (Fig 119).   Cuspate sand bars
frequently tie offshore bedrock islands to the land (Fig
120)[96].
   The mouth of each inlet or bay will be shut off by a
bar, but this will be breached by a tidal entrance deflected in the direction of the offshore current.   Most
such entrances have a double-headed delta - the outer
delta will be built by the ebb tide and stream flow, the
inner delta built by flood tides that sweep into the bay
(Fig 121A).   The outer delta is smoothed and reduced by the
work of offshore currents and storms.
   Hurricanes will also deposit inner deltas where they
break the bar.   At normal times these are not connected
with the ocean, although they are opposite a low place in
the outer sand reef (Fig 121B).   The water crosses these
low places only during great storms, when onshore winds
pile up the waves.

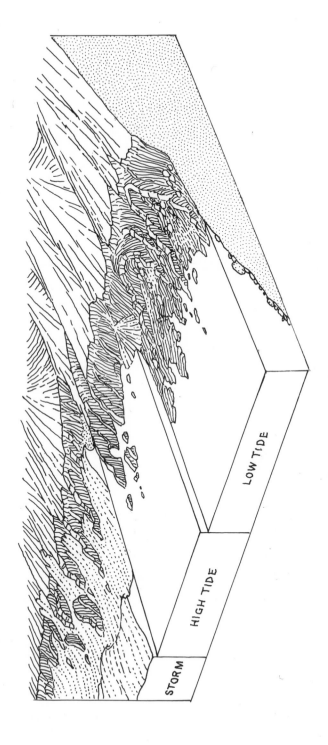

Figure 113. Diagram showing characteristic features of a cliffed shoreline of submergence (from Davis, 1908, fig 1, p 25).

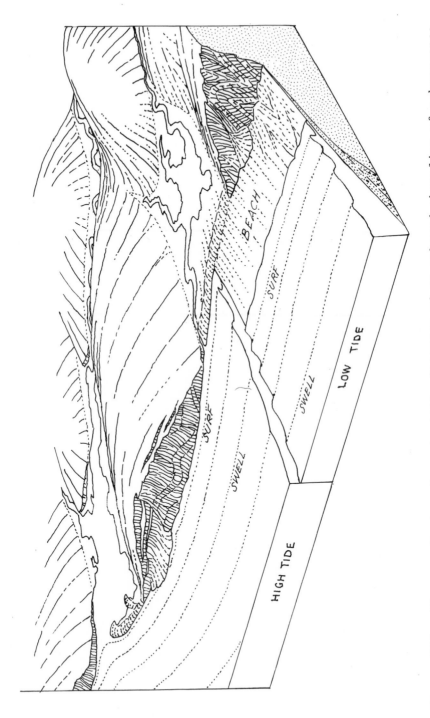

Figure 114. Diagram showing characteristic features of an embayed shoreline of submergence (from Davis, 1908, fig 2, p 25).

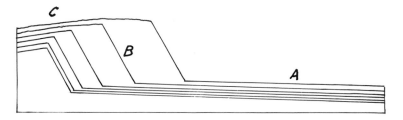

Figure 115. Profiles comparing rates of lowering or back-
wearing of: beach, A; sea-cliff, B; and land surface, C.
Spacing of the lines represent equal time (from a
blackboard sketch). (see note [93] re Fig. 111).

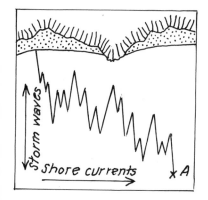 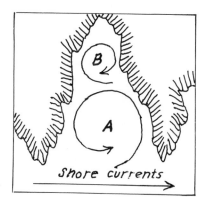

Figure 116. (left) Plan showing movement of material from
beach to final resting place in deep water, A, pro-
pelled by storm waves and shore currents (from a blackboard sketch).

Figure 117. (right) Plan showing production of backset
eddy, A; and foreset eddy, B; in a bay head by action
of shore currents (from a blackboard sketch).

    The processes of submergence of shorelines just
described produce varied forms on the adjacent land, a few
samples of which are shown in the four views shown in
Fig 122.
    At a later stage, inequalities of the coastline are
gradually erased, and the offshore current is free to sweep
the shore. The rivers will be driven in the direction of
the current, which will deflect their tidal entrance. On
the cliffed headlands the valleys will be left hanging
above the beach, forming marine hanging valleys, or
valleuses – a name derived from their typical occurrence on
the chalk cliffs on the coast of Normandy(Fig 123)[97].
    Full maturity is attained when all the bays are cut
away. By now, not only are the cliffed promontories subject to erosion, but also the deltas in the bays. Besides,

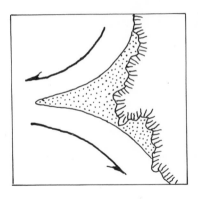

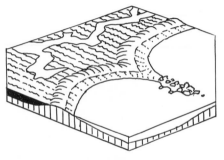

Figure 118. Formation of a cuspate foreland by currents (from a blackboard sketch).

Figure 119. Dungeness on coast of southeastern England, an example of a cuspate foreland (from a blackboard sketch).

Figure 120. Cuspate sand bars formed by shore currents, which tie bedrock islands to the land. In the right-hand diagram, the island has been nearly destroyed by marine erosion, leaving only a rock reef (from Davis, 1912, figs 197 and 198).

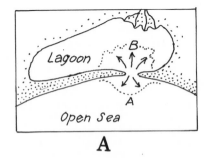

A

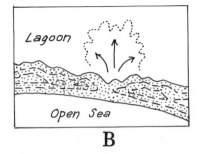

B

Figure 121. Deltas adjacent to sand reefs: A) Double-headed delta at lagoon entrance; a) formed by the stream that enters the lagoon, b) formed by the tide. B) Delta in a lagoon, formed by a hurricane (from a blackboard sketch).

Figure 122. Varied landforms produced along shorelines of
    submergence: A) Port Lyttleton, a drowned valley in
    the western part of the Banks Peninsula of New Zealand.
    B) Sea cliffs on north shore of Banks Peninsula, New
    Zealand. C) A steeply plunging coast; the Riviera di
    Levante southeast of Genoa, Italy. D) A filled-in bay
    head; west coast of St. Lucia, in the Lesser Antilles
    (from Davis, 1928, figs 61, 62, and 97; Davis, 1912,
    fig 2).

The marine cycle

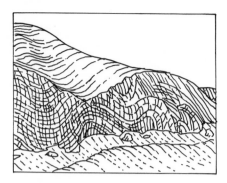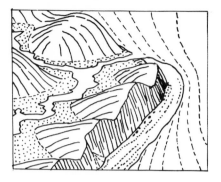

Figure 123. (left) Marine hanging valley or valleuse (from
Davis, 1908, fig 4b, pl 28).

Figure 124. (right) Sea cliffs of decreasing height (from
Davis, 1908, fig 4a, pl 28).

the cliffs have now been cut back so far that some of them
will have decreasing height (Fig 124), where the land
slopes to some interior valley. Eventually, the waves will
cut into such valleys, which originally were behind the
shorelines.
    Old age begins when all the shoreline is retreating
relatively evenly, with the original bays and promontories
destroyed[98]. With late maturity and old age, erosion by
the sea slows steadily, and finally becomes imperceptible.
It is uncertain how long such erosion can remain effective.
However, perhaps if a whole continent remained at stillstand long enough, all of it could eventually be reduced by
the waves. The maximum known height of sea cliffs is about
1000 feet (in Ireland and Normandy); the maximum width of
wave-cut terraces - the baseleveled surface of marine
erosion - has never been recorded[99].

Region of San Francisco Bay

As shown earlier (see section on Erosion Features of Folded
Mountains) the Coast Ranges of California consist of many
sub-parallel ranges, which are blocks of disordered structure, trending northwest-southeast, of moderate height and
submature or mature dissection. Between the uplifted
blocks are more depressed blocks, producing intermont
troughs. The troughs are aggraded into intermont plains,
the detritus being supplied by the dissection of the uplifted blocks.
    Here and there the larger rivers pursue transverse
courses through the ranges; for example, the Russian River
to the north and the Pajaro River south of San Francisco.
One such transverse river heads in the great aggraded plain
of the Valley of California, and received the Sacramento
River from the north and the San Joaquin River from the
south-southeast. It crossed the inner range in Carquinez
Gorge, then flowed across a well-aggraded intermont plain,
and finally reached the Pacific Ocean through the Golden
Gorge in the outermost range.

The region was slowly submerged some 400 feet[100]. The
Golden Gorge was submerged to form the Golden Gate; the
intermont plain was broadly submerged in the confluent bays
of San Francisco and San Pablo, but non-submerged parts of
the plain extend north to Petaluma and Napa, and south to
San Jose. The Carquinez Gorge was submerged in the narrow
Carquinez Strait; part of the inner plain, or Great Valley,
was submerged over a small area in Suisun Bay.

The shorelines of the bays are simple, because they lie
on smooth plains; shorelines of the ranges are irregular,
because the slopes of the ranges were well dissected before
submergence; witness the beautiful embayments and points of
Tamalpais.

Since submergence, the exterior shoreline outside the
Golden Gate has been much cliffed by the vigorous ocean
waves, and Bolinas Bay has been closed by a beach. The
interior shoreline had been little cliffed by the feeble
waves of the bays, but the Tamalpais spur ends have been
cliffed a little; the bay heads are already encroached by
delta flats. The largest of such deltas are the confluent
and marshy deltas of the Sacramento and San Joaquin Rivers
in Suisun Bay.

Shorelines of emergence

On a long, smooth coast of a shoreline of emergence, the
tides do important work, and these now deserve discussion.
In a bay head, for 6 hours and 25 minutes the water will
rise slowly, and then for 6 hours and 25 minutes, it will
fall slowly. In some bays there is a great tidal range,
but there are no currents. On an open coast tides are
chiefly manifested by tidal currents. A tidal current is
a vector of the forces that cause the tides; the tidal
forces act from east to west, and the relation of these
forces to the trend of the shore on which they impinge,
determines the dimensions of the tidal currents.

We shall assume a rapid emergence of a shoreline, with
few if any pauses. The newly uplifted shore was the former
sea-bottom; the shore is straight and even, and the sea-
bottom is very shoal. Behind the shore is a young coastal
plain (see section on First-cycle coastal plains).

As the waves advance on the shore they will reach a
point where the wave base touches the shoal bottom. Here,
sediment is stirred up by faint oscillations. If there are
tidal currents at work, they will move the loosened
material.

Where the waves break, there will be a lifting and
shaking of the sands, and these will be delivered to the
currents, which will gradually carry the finer material
out below wave base. The coarser sands will accumulate
inside the zone where the surf breaks. This heaping of
coarser sands is the beginning of a bar or sand-reef
(Fig 125A). The bar will be steepened toward the seaward
side by excavation by the waves. The bar will finally rise
above the waves as a sand-reef[101], by action of storm waves
and the wind. Behind the sand-reef is a lagoon, which will
gradually be filled in and become a tidal marsh (Fig 125B).

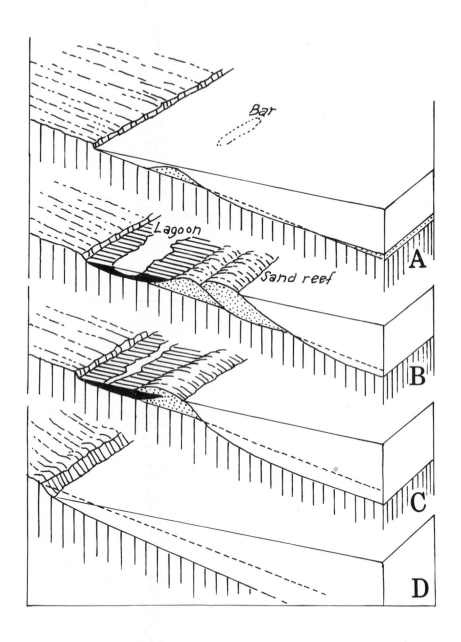

Figure 125. Stages in development of a shoreline of emergence (from Davis, 1898, fig 223).

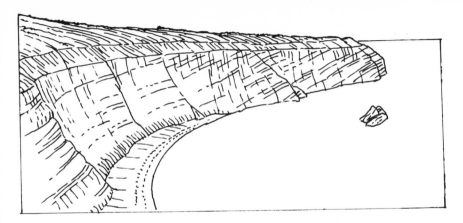

A

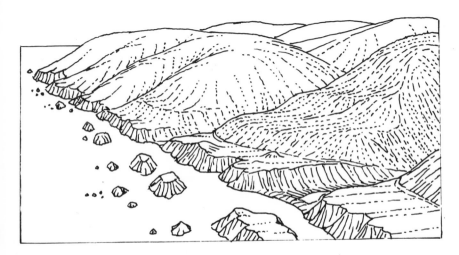

B

Figure 126. Uplifted marine terraces on the coast of California: A) A marine terrace 100 feet above sea level, truncating tilted Miocene strata; near Laguna south of Los Angeles. B) A strong lower marine terrace and two higher fainter terraces; on coast near Cape Vizcaino north of San Francisco (from Davis, 1933, figs 24, 26).

Exceptional additions to the outer face of the sand-reef will be made by rare, enormous storms (the greatest storm in a thousand years). Development of offshore sand-reefs marks the beginning of maturity along the coast.

Eventually, the waves will expend all the energy they are able on the sand-reef, and will begin to deepen the bottom in front. This will permit them greater freedom of motion, and they will begin to advance on the sand-reef itself. Material is tossed across the bar into the lagoon behind, partly burying the tidal marsh deposits (Fig 125C). As the bar is eroded, tidal marsh mud will crop out on its face - a feature which can be observed in some of the New Jersey sand-reefs.

The shoreline is now in full retreat; the lagoon is soon destroyed, and the mainland is undercut in a bluff (Fig 125D). Continued erosion, as in the old age stage of a shoreline of submergence, will result in continued destruction of the mainland, once far from the shoreline.

Emerged shorelines, raised far higher than in the simple example just described, occur along many parts of the coast of California. Marine terraces standing 100 feet above present sea level are common, and on some coasts flights of higher, less well-preserved terraces occur, to a number of 10 or more. The lower of these terraces are related to the abstraction and addition of sea water during the ice ages; the higher ones may have other causes (Fig 126).

Coral Reefs

Observable features

A special form of shoreline is the coral reef[102], composed of
masses of lime-secreting organisms whose skeletons are
built up into wave-resistant frames, the whole forming a
community adapted to life in shallow, agitated sea-waters.
Dominant elements in such communities are corals, although
other sessile lime-secreting animals, such as sponges and
bryozoans, play lesser or greater roles. Important
constituents are also the lime-secreting algae, or nulli-
pores, whose lime serves chiefly as a binder for the
skeletons of the animals. Corals can live at various
depths, down to fairly deep water, and in various tempera-
tures from warm to cool, but the massive, reef-building
types are virtually restricted to depths of 25 fathoms or
less, and to warm, tropical seas. Corals and coral reefs
require well-aerated, clear waters for their existence.
They can survive neither in stagnant waters nor in waters
which contain much suspended detritus. Sandy and muddy
waters, such as those shed off actively eroded lands, will
smother and kill the reefs.

The chief material secreted by the reef-building
organisms is calcium carbonate, which is super-saturated in
warm, tropical waters. However, if the reef carbonates are
exposed to sea water for any length of time, they are
altered to dolomite by the addition of magnesium; thus,
even very recently deposited reef rock is commonly
dolomitic.

Morphologically, coral reefs are divisible into three
general kinds - fringing reefs which hug the land, barrier
reefs which are farther out and separated from the land by
a lagoon, and atolls in which the reefs form a ring which
surrounds a lagoon with no land area. The three kinds are
gradational, with transitional types such as almost-atolls,
whose central lagoons contain a few rock islets.

In barrier reefs, the most instructive of the three kinds, the reef forms a massive frame resistant to wave attack, whose outer face slopes gently or steeply into deep or very deep water. Presumably the outer slopes consist largely of reef detritus, which in the steeply sloping parts is principally talus. Behind the barrier reefs are lagoons, half a mile to as much as 10 miles broad. Reef slope, reef, and lagoon thus form a terrace with respect to the adjacent land. Lagoons are commonly 20 to 40 fathoms deep, but exceed 60 fathoms in a few places. The lagoons, protected from the open oceans by the reefs, are areas of relatively quiet water, which are the domain of various specialized organisms, such as delicately branched corals. Patch reefs, and even some fringing reefs occur in the lagoons, but parts are strewn with blocks of dead coral rock, evidently thrown in from the barrier reefs by storm waves.

In the lagoon there is a constant filling and leveling by the waves and currents, which smooth the detrital material delivered to the lagoons - partly carbonate rocks from the reef itself washed in by storm waves, partly inorganic detritus brought down by streams from the adjacent lands, if any occur. Anti-leveling processes seem rather implausible, and especially solution of previously deposited reef material, as waters in low latitudes are supersaturated with lime, and lime is in the process of precipitation on the barrier reefs not far away. Nevertheless, leveling is not complete, as the lagoons contain occasional holes 60 fathoms or more deep[103].

Constant delivery of materials from the reef and adjacent lands would suggest that eventually the lagoons would fill up and become low-lying land. That the lagoons are maintained as water areas seems to require an interposition of additional processes.

Coral reefs commonly occur on oceanic islands, most of which are volcanoes of mafic lavas that were erupted from and built up on the sea floor. Only a few reefs, as along the northern Australian coast, near New Guinea, and in New Caledonia, have been built along the edges of continents, or along blocks of continental crust. The mafic volcanism of the oceanic islands, while incidental, requires some consideration in interpreting the coral reef problem.

Does the sequence from fringing reefs to barrier reefs to atolls suggest some evolutionary process, and if so, what process? Search for this process is the crux of the coral reef problem. The observed features of coral reefs can be recorded in great detail, but their origin and their past history lie within the realms of historical geology and interpretative geomorphology; this history cannot be observed, but deductions can be made from the existing nature of the reefs, and of the associated lands and seas.

Investigations and interpretations of coral reefs

Coral reefs of the tropical seas have excited the interest and imagination of explorers for hundreds of years, and many fanciful explanations of them were made in the reports of the early expeditions.

Darwin's theory

The first significant interpretation of coral reefs was
made by Charles Darwin, as a result of his observations
while still a young man, during the voyage of the Beagle
in 1832 to 1836, which visited many of the coral islands
of the Pacific. Darwin's interpretations were presented
briefly in 1837, and in book form in 1842, with many subsequent reprintings.

Darwin saw in the fringing reefs, barrier reefs, and
atolls a sequential development of forms, the earlier
evolving into the later, as a result of <u>subsidence</u> of the
reef-girt islands, and consequent upgrowth of the reefs
(Figs 127, 128).

Nevertheless, the morphology of the forms to be expected as a result of subsidence and upgrowth were in-
completely analyzed. An approach to such an analysis was
made by James Dwight Dana, who also served as a young man
as naturalist for the United States Expedition under
Captain Wilkes, about a decade after Darwin's work. This
expedition did important work in the Antarctic and along
the Pacific Coast of North America, and it also visited
many of the Pacific coral islands. Dana deduced that if
these islands had subsided, their valleys should be <u>drowned</u>
and their mouths <u>embayed</u>; he found confirmatory evidence
for these features on Tahiti (Fig 129).

Daly's glacial control theory

Early in the present century the effects of the Pleistocene
ice ages on conditions in the oceans began to be appre-
ciated - the extensive cooling of the ocean waters, and the
lowering of sea level as a result of locking up of water in
the continental icecaps. Application of these concepts to
the interpretation of coral reefs was made by R.A. Daly in
a series of papers between 1910 and 1919[104].

Daly postulated that any coral reefs which might have
existed earlier were killed during the glacial period,
partly because of cooling of the ocean waters, and partly
because of turbid conditions created on the submarine banks
as a result of lowered sea level. With the reef barriers
rendered ineffective, marine erosion had full play over the
former submarine banks, now nearly or completely exposed to
view as a result of lowered sea level, and they were thus
planed off to depths of about 30 fathoms (Fig 130). The
modern reefs - fringing, barrier, and atoll - were believed
to be post-glacial features which grew on the planed-off
surfaces thus created, during a subsequent rise in sea
level. An alleged uniformity everywhere of lagoon floors
at about the stated depth was cited as evidence of this
low-level planation, and was considered to be a fatal ob-
jection to Darwin's belief that they were filled during
subsidence.

Unquestionably, the Pleistocene ice ages, by cooling
the water and lowering the sea level, had a profound effect
on conditions in the coral seas, which must be considered
in any interpretation of coral reefs, but some of Daly's
assumptions seem implausible: a) Prolonged stability was

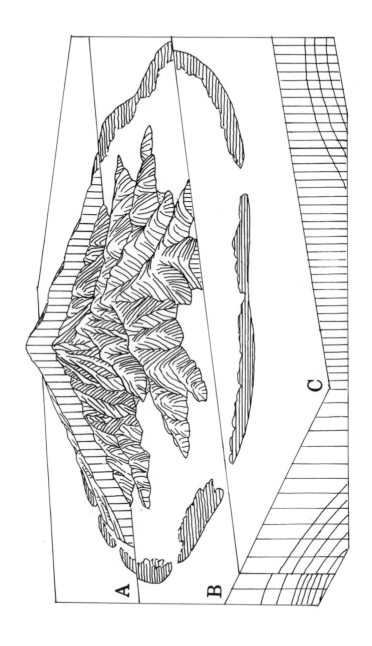

Figure 127. Three-stage diagram showing a subsiding island in a still-standing ocean. A fringing reef in the background block is converted into a barrier reef in the middle block, and into an atoll in the foreground block (from Davis, 1928, fig 20).

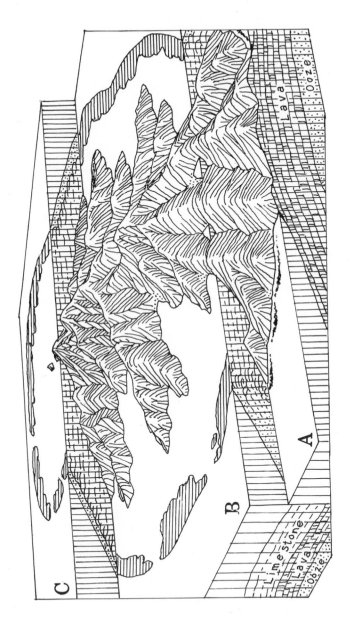

Figure 128. Three-stage diagram of a still-standing island in a rising ocean. A fringing reef in the foreground block grows up into a barrier reef in the middle block and an almost-atoll in the background block. The forms produced are the same as those in Fig 127, although the cause of the change in the water level differs (from Davis, 1928, fig. 21).

Figure 129. Northern coast of Tahiti, showing embayed and
delta-filled valleys, separated by spurs that are
cliffed at their seaward ends (from Davis, 1928).

assumed in the coral seas, extending far back into
Tertiary times and before, whereas these seas are tectonically active regions. b) The volcanic islands or other
foundations were supposed to range from Tertiary to Precambrian, hence subject to prolonged weathering and
erosion, so that they were already very susceptible to the
forces of marine erosion when they were attacked during the
glacial period; later evidence indicates the extreme youth
of most of the oceanic islands. c) Little effort was made
to account for reef growth before the ice ages, and some of
the postulates of the theory seem predicated on their non-
existence.

Observations by Davis

An understanding of the problem would be assisted by a more
rigorous morphological study.
     The approach to the problem should be morphological, in
terms of the land and water forms that were produced, and
their relation to the various opposing theories. With this
viewpoint, what matters in the reefs is not their biology
and ecology, but the nature of their form, and what is
significant in the volcanic foundation of the reefs is not
their petrography or petrogenesis, but whether the rocks
of the foundation are strong enough to produce cliffs and
mountains, or weak enough to be carved into valleys or to
be planed off by the sea. Morphological history of the
lands is as significant as that of the reefs themselves,

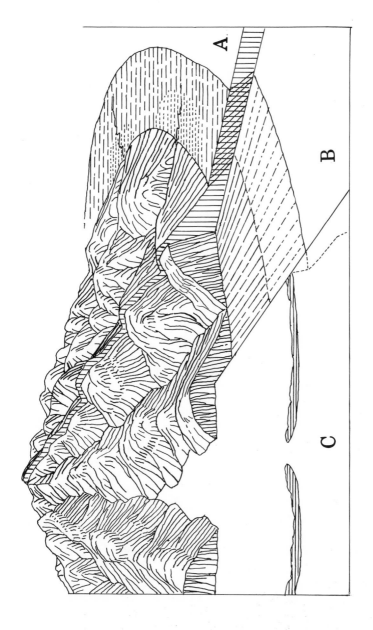

Figure 130. Sequential diagram showing an epoch, B, of low-level abrasion on a still-standing island, followed by a rise in water level and formation of a barrier reef according to the Daly glacial-control theory. Sequence proceeds from A to C (from Davis, 1928, fig 40).

because of the indications which they provide of changes in
sea level, and other history.

The usual methods were applied - deduce the consequences
of the respective theories and confront these consequences
with the facts. The resulting interpretations are in
agreement with those of Darwin - that coral reefs grew on
subsiding foundations, and that the reefs and their
associated features have an evolutionary sequence of forms
resulting from subsidence. This general history is modified, however, by the obvious effects of the ice ages, the
cooling of the oceans and the lowering of sea level,
especially in the marginal belts of the coral seas[105].

Within the coral seas themselves, islands can be
grouped into a sequence of forms, believed to have resulted
from evolution under conditions of steady or intermittent
subsidence.

1. Non-submerged, reefless islands, with simple
    cliffed shorelines.
2. Partly submerged islands with fringing reefs,
    having partly submerged cliffs and embayed shore-
    lines (youthful stage).
3. Further submerged islands with barrier reefs,
    having cliffed and embayed shorelines similar to
    those of the preceding (Figs 131 and 132).
4. Islands with barrier reefs, whose earlier cut
    cliffs have been submerged, and whose interbay
    spurs taper into widening barrier reef lagoons
    (mature stage).
5. Islands so submerged that only a few islets project
    in the lagoon; the barrier reef is an almost-atoll
    (late stage).
6. Atolls, in which the central island has disappeared
    (stage of extinction).

An example of this sequence of forms occurs in the
Society Islands of the south-central Pacific, a chain about
200 miles long. Mahetia, the southeasternmost island, is
a young and reefless volcanic peak. Tahiti, next to the
northwest, is deeply dissected, with filled-in drowned
valleys, low plunging cliffs at the ends of the intervalley
spurs, and a barrier reef (Fig 129). Succeeding islands of
the chain to the northwest are deeply dissected, without
cliffed spur ends, but with prominent barrier reefs (Figs
131, 132). Beyond are almost-atolls, then atolls.

This sequence of forms suggests southeastward progression of volcanic activity, probably along a fracture zone
on the ocean floor, each volcano in turn having been built
up above sea level, then becoming extinct and gradually
submerged. A similar southeastward progression of volcanic
activity is well authenticated in the Hawaiian Islands, but
these islands are in the marginal belts of the coral seas,
so that the associated reef forms are less well displayed.

Many lines of inference, besides those based on the
reefs themselves, suggest development of these and other
coastal oceanic islands in a region of subsidence. The
islands were eroded dominantly by streams, which have
scored them with deep valleys, and they have only been subjected to marine erosion during the early stages. The
early, minor sea cliffs are soon lost by submergence, and

Figure 131. Sharply dissected volcanic rocks of Raitea, Society Islands, northwest of Tahiti (from Davis, 1928, fig 127).

Figure 132. Borabora, Society Islands, northwest of Tahiti, looking north; volcanic peaks with morning clouds above (from Davis, 1928, fig 135).

the lower ends of the stream valleys are drowned, as
observed long ago by Dana, producing an embayed coastline.
Stream erosion removed large quantities of detritus from
the islands which were delivered to the lagoons, which
should have filled up, but this detritus has disappeared;
the only manner in which the detritus could have been
accommodated would have been by subsidence. The islands of
the almost-atolls are little peaks with sloping sides,
suggesting nearly submerged mountain tops; under the Murray
(1880, 1887) theory of stillstand, they should have been
worn down to low relief; under the Daly theory of glacial
control they should have been strongly cliffed, or should
be merely stacks and pinnacles.

If the reefs formed on subsiding volcanic islands, the
reef and lagoonal sediments should lie on their volcanic
foundation with a surface of unconformity, indicating that
the volcanic bedrock had been a land surface that underwent
subaerial erosion prior to reef deposition (Fig 133). If
the reefs had been formed on a stable volcanic foundation,
the rock beneath them had always been beneath the sea, and
there should be more or less gradation from lavas and volcanic sediments into reef carbonates. Test of this
criterion in most islands in the open ocean is not possible,
in the absence of drill data[106], as the contacts are out
of sight below sealevel, but it is visible in upraised
reefs of the tectonically active southwestern Pacific,
where such an unconformity and surface of erosion is
visible in many places; during formation of these reefs,
their foundation subsided, even though they were uplifted
later.

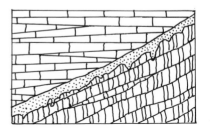

Figure 133. Unconformity
between reef rock and its
volcanic foundation, indicating an intervening period
of subaerial erosion (from a
blackboard sketch).

Presumably the islands of the coral seas, like those
of the marginal belts, were subjected to lowered sea level
during the glacial periods, as well as to a subsequent
rise in sea level. Thus, their reefs were partly or wholly
emergent during this period. Nevertheless, as the tropical
reefs formed strong, massive barriers, even when emergent,
they protected the island shores from wave attack. It
seems implausible that such massive barriers originated
wholly during the relatively brief time since the last ice
age; whatever reefs grew during this time were built on the
even greater barriers whose origins must have been farther
back in the past.

PART 4 ACCIDENTS

The Volcanic Accident

In geomorphology, an accident is an interruption of the
erosional regime of the region by some new factor, which
modifies the orderly, sequential development of the landforms, yet does not change such gross relations as altitude
or structure. Within certain limits of plausibility,
accidents can occur in a region of any structural form, at
any place, at any time, and at any stage of erosional
development.

Here, we shall consider two such geomorphological
accidents - those produced by volcanoes, and those produced
by glaciation.

Volcanic accidents are susceptible to the usual methods
of geomorphological analysis; a description of a volcanic
region should include statements regarding: a) the nature
of the region before the eruption, b) the topographic forms
produced by the eruption, and c) the modifications of these
forms by other processes after the eruptions have ceased.

Landforms created by volcanic accidents

Lavas and associated gases, compelled to rise from a deep
source, make their way upward through the crust in chimneys:
some eruptions also issue from fissures and their lavas
overspread wide areas. Volcanic cones have relatively
gentle declivities if built up chiefly by fluid lavas, and
steeper slopes formed of viscous flows, ash, or cinders.
Fine volcanic dust or ash may be carried high into the
atmosphere during the eruptions, and will blanket wide
areas, or fall into the sea; although such ash falls will
commonly kill the vegetation, they are ordinarily so thin
that they little modify the landscape.

Rains - some of which are provoked by the eruptions -
will wash ash, cinders, and coarser fragments down the
slopes of the cones, and reduce their declivity while

broadening their basal diameters. During dormant periods, consequent streams on the slopes of the cones will score them with radial valleys, damaging their constructional forms. Later eruptions may repair this damage, and some cones will be heaped up higher than before.

At the summit of the cone, a crater is kept open by gas explosions when the volcano is active. Huge summit cavities, or <u>calderas</u>, are sometimes produced by great eruptions or engulfment (Fig 134). Their walls break across previously built-up structures, whereas craters have built-up rims; however, there are many intermediate forms. After the eruptions cease, both craters and calderas are subject to modifications by erosion. Crater Lake in southern Oregon is a superb example of a caldera, formed by the collapse of one of the great volcanoes of the Cascade Range.

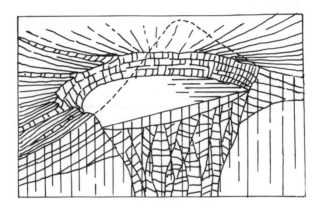

Figure 134.   A caldera (from Davis, 1898, fig 131).

In studying the geomorphic effects of a volcanic accident we can begin with a simple case - an eruption in a region with little previous volcanic history, from a single vent or closely related vents, followed by extinction of volcanic activity. Examples of such simple volcanic accidents occur in many parts of the Rocky Mountains, and in the Rhenish Highlands of western Germany.

In such a region, which is undergoing erosion, a volcanic cone may break through at any stage in the cycle, and in any topographic situation. Let us assume, however, that the eruption occurs when the region is maturely eroded, and that the vent emerges on an interfluve, above existing drainage levels (Fig 135A). The vent will build up a cone of ash and lava, in the manner indicated above, and from it or its flanks lava will flow. Lava will run down pre-existing stream valleys as far as their current will carry them before they solidify. While flowing, the lava surfaces are more or less congealed; their cross-profiles are generally convex (Figs 135B and C).

Where the lavas fill valleys, volcanic lakes will be impounded upstream above the lava dams. Downstream, the

former drainage of the now-filled valleys will be deflected
sidewise from their original courses against the country
rock of the adjacent uplands; new drainage will take its
course on one or both sides of the lava filling (Figs
135B, C). The deflected drainage will cut new valleys in
the country rock, which will eventually drain the lakes
upstream (Fig 135D). When the lavas congeal they form
rocks that are more resistant to erosion than most of the
original country rock. Thus, as degradation proceeds, the
lava that filled the original valleys will stand as ribbon-
like mesas between the new valleys (Figs 135E, F). These
mesas may persist long after other traces of the pre-
volcanic topography have disappeared, and will afford the
only indications of its nature.

A new consequent drainage will form on the surface of
the volcanic cone and its adjacent lava field, whose
streams will radiate down the slopes of the cone and will
follow the surfaces of the lavas farther out (Fig 135D).
This radial drainage will persist after the constructional
volcanic surfaces have been destroyed, and even after most
of the materials resulting from the volcanism have been
removed (Fig 135G). In the later stages, consequent
drainage resulting from the volcanism will be superimposed
on the original bedrock, which necessarily has a very
different structure.

After the eruptions have ceased and the volcanic field
has become extinct, erosional processes again dominate, and
most of the original constructional volcanic materials are
removed. Then, the deeper-seated intrusive products of
the volcanic accident will come into view (Fig 135H). If
the neck of the volcano was filled with material more
resistant than its surroundings it will project as a peak
or pinnacle. Dikes of volcanic rock which were intruded
radially from the vent are also likely to be of material
more resistant than their surroundings, and when eroded
will project in wall-like ridges.

Examples of volcanic accidents

Mount Shasta in northern California has the appearance of
a perfect cone, but is actually maturely dissected and
scored by gigantic ravines. Around its base are meadow
lands that were once lakes dammed by lava barriers; these
have since been filled by sediment, and have been partly
drained by outflowing streams.

The Cantal in the Central Plateau of France is a more
deeply dissected volcano about 40 miles in diameter. Its
present height has been greatly reduced from the original
by later erosion, to judge by the slant of the lavas on its
flanks, and it has been cut to pieces by radiating valleys.
A railroad and highway enter the region from the southwest
by one of these valleys, cross the volcanic center through
a low pass, and leave through another valley on the northeast.

More advanced stages in the erosion of simple volcanic
fields are abundantly displayed in the Colorado Plateau of
northeastern Arizona and northwestern New Mexico, where

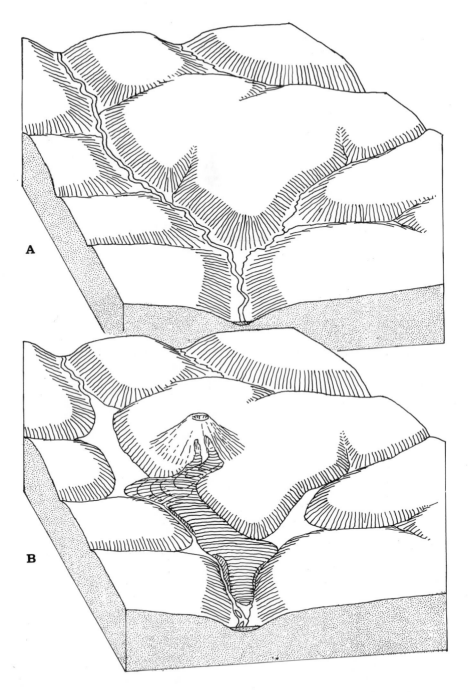

Figure 135. A series of eight diagrams (A through H)
illustrating the sequence of events during a simple
volcanic accident (from Davis, 1912, figs 127-134).

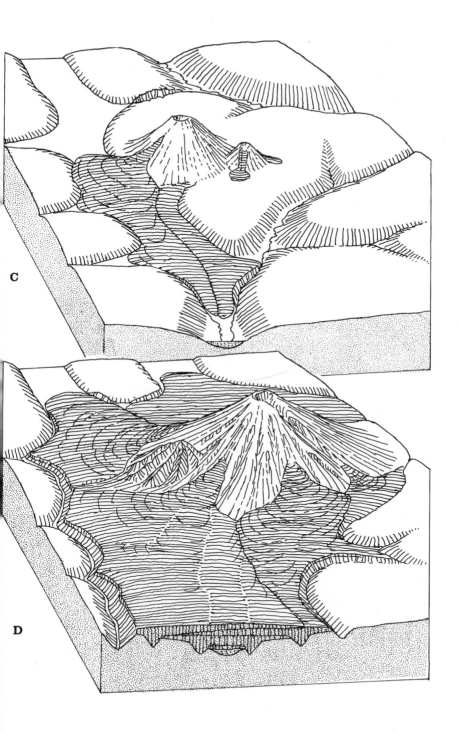

C

D

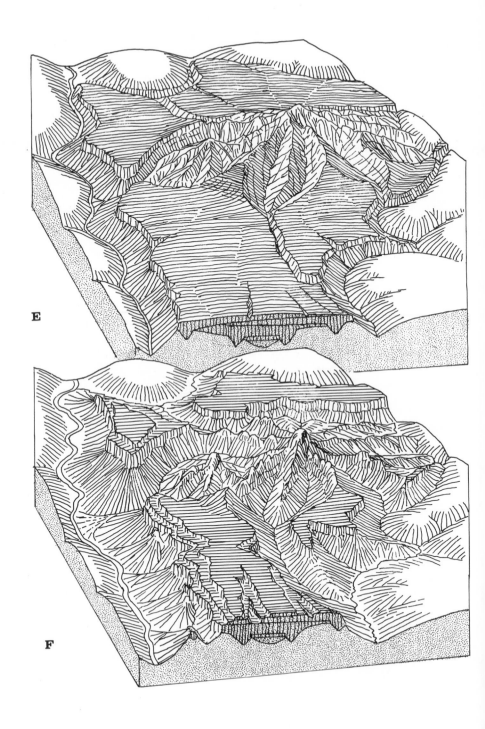

E

F

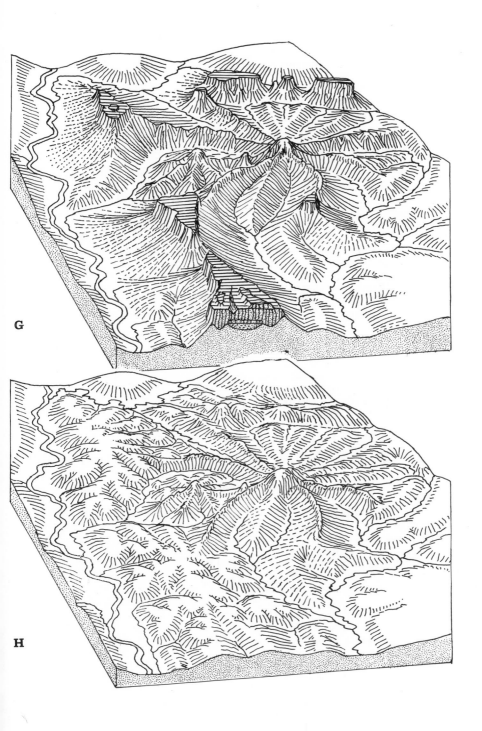

G

H

they are eroded into striking topographic forms under the
influence of the arid climate. Volcanic necks, such as
Shiprock, form prominent spires and peaks, some projecting
more than a thousand feet above their surroundings. Somewhat more subdued forms are created in the former volcanic
area of the Monteregian Hills of southeastern Canada,
where the climate is humid and the eruptions were older,
but even here the necks form prominent hills or low mountains, one of which has given its name to the city of
Montreal.

An interesting volcanic district

The following is a geographic   (geomorphic) description of
a well-known area:

1) A mountainous background, trending northwest-
southeast, with a piedmont lowland belt 20 miles wide
between the mountains and the sea to the southwest, is
crossed by consequent streams draining from the mountains
in the background to the sea in the foreground (Figs
136, 137).

2) A chain of four contiguous volcanoes, each 10 to
15 miles in diameter at the base and 2000 to 3000 feet
high, is built on the piedmont lowland; the first and
fourth lie against the mountains, while the second and
third stand farther forward, enclosing a lowland behind
them. The third and fourth volcanoes are farther apart.
All the cones are confluent at their bases, separated by
relatively low saddles.

3) The mountain streams back of the volcanoes find
their courses to the sea obstructed. They flood the enclosed lowland, forming a lake; the lake rises, seeking an
outlet across the lowest intervolcanic saddle or col,
which lies between the third and fourth volcanoes, by
reason of the fact that they are farther apart than the
rest. The lake outlet cuts down a flat-floored valley,
drains the lake, and all the blockaded streams join the
outlet stream, which thus becomes a good-sized trunk river,
and pursues a meandering course from the cross-col valley
to the sea.

4) Many small streams cut radial ravines in the volcanic slopes; lateral streamlets fray out the inter-ravine
spurs, especially where the spur ends are cut off in the
low bluffs on the sides of the cross-col valley, which thus
comes to be bordered by many hills - 30 or 40 on each side
or about 70 in all. The detritus from the volcanic ravines
and from the background mountains is all swept along by the
trunk river, and forms a delta at the seashore.

5) This volcanic district comes to be occupied by
barbarous tribes, who settle in many villages. One of
these villages was located in the center of the third
volcano, another in the cross-col valley, and several
others at scattered locations.

6) As the barbarians advance toward civilization, one
of the villages outgrows the others, and becomes (in the
course of 2000 years or more) an important city. This was
the village located at the cross-col valley which, by

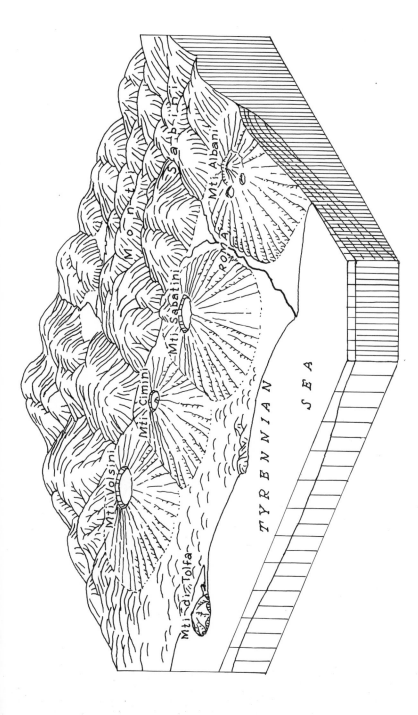

Figure 136. Diagram of the environs of Rome, showing the four volcanoes. Block is oriented northwest-southeast, and the long dimension is about 100 miles (from Davis, 1912, fig 137).

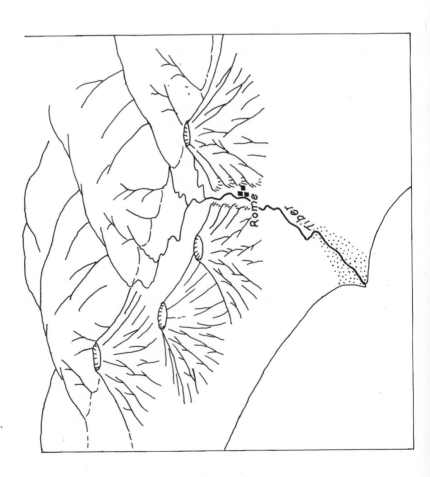

Figure 137. Davis's blackboard sketch of the same area as that shown in Figure 136 (as copied by P.B. King). The great exaggeration of vertical scale makes identification of the area difficult.

reason of its natural environment, had certain advantages
for growth.

7) The city is <u>Rome</u>. Its Seven Hills are but a part of
the frayed-out system on volcanic remnants which flank the
cross-col valley. The proverb 'all roads lead to Rome'
expresses the advantages of its natural location, where
all roads must necessarily converge at the point of passage
from the sea to the plains, and the mountain behind.

Compare this geomorphic description of the city with
the description of the city as you have hitherto known it.
Which description gives the best idea of its location?[107]

Volcanic plains and plateaus

In some parts of the world lava has poured out in such
volume that it has spread far and wide over large areas.
The lavas were so fluid at the time of the eruptions that
they flowed into the lowest ground available, spreading out
and solidifying to form a lake-like plain. After eruption,
they have been subjected to deformation and erosion, in the
same manner as the other rocks of the earth's crust.

The Columbia Plateau and Snake River Plain in the
interiors of Washington, Oregon, and Idaho have been
covered by sheets of basaltic lava of great thickness and
extent. They bury the lower ground and lie against the
bordering mountains, rounding off the spur ends and
entering the valleys. In some places they have been little
modified by subsequent erosion; in others they are dissected, and in still others they are deformed into folds
and fault blocks, and have become mountainous.

A large part of the Deccan, or southern plateau of
India, is lava-sheeted. Originally these sheets extended
westward into the present site of the Arabian Sea, but this
part is fractured and depressed below sea level. The
eastern part has been upheaved into a plateau, whose
western edge, next to the region of subsidence, projects
as a scarp that has been moderately dissected.

Inner Brazil has large lava-covered areas. Small gas
cavities near the surface of the lava sheets are slowly
filled with silica in the form of agate. As the lava sheet
weathers, the agates, more resistant, remain as pebbles and
cobbles, which are gathered and shipped to Germany, where
they are polished and sold.

The Glacial Accident

Like volcanism, glaciation is an accident which interrupts
normal processes without changing the gross structure,
altitude, or other environmental controls of the region.
There are, however, restrictions as to the region that
could be affected by the glacial accident; such an accident
is unlikely in the tropics, and less likely in lower
country than in higher mountains, where only slight climatic changes will increase retention of snowfall.
Glaciation in lower country, producing continental icecaps,
requires prolonged and severe refrigeration of a large
part of the earth. The present discussion deals solely
with mountain (= alpine) glaciation.

Can glaciers excavate?

During the latter part of the Nineteenth Century many geomorphologists, especially those working in the Alps,
strangely refused to believe that present or past glaciers
could have much erosive power, or that glaciers have
greatly modified the mountain landscape. During the same
period, J.D. Whitney and Clarence King derided John Muir's
contention that the Yosemite and other valleys in the
Sierra Nevada had been sculptured by glaciers. Many in-
genious arguments were presented by these opponents of
glacial erosion, none of which will withstand close
scrutiny. Landforms in a region that has been subjected to
alpine glaciation are so different from those in a region
that has been subjected to normal erosion that some very
unusual processes must clearly have been at work there,
which could scarcely have been other than glaciation.

Habits of glaciers

Very commonly, mountain glaciation occurs in a region which
previously had rounded forms and an insequent drainage,
shaped by normal processes of erosion (Figs 138A, 139A).
When the period of glaciation begins, ice which was derived
from snowfall accumulates in small patches at the upper
ends of the valleys. These ice patches increase in size
from winter to winter, and those in adjacent valleys
eventually coalesce into an ice field, from which glaciers
move downslope until they reach a level where their lower
ends melt away.
        Glaciers resemble streams of water and lava in that
they move downslope, and like lava they will seek valleys
that had previously been cut by streams (Figs 138B, 139B).
Unlike lava, and like streams of water, glaciers will much
erode these valleys. Nevertheless, glacial ice is vastly
more sluggish than running water, being about 20 000 times
less nimble[108]. Like streams, glaciers move more rapidly in
the center than at the sides, but unlike water streams they
have little centrifugal force in rounding valley bends; on
the other hand, inertial forces will cause them to cut more
deeply on the inner rather than the outer sides of the
bends. Thus, glaciers tend to straighten previously
sinuous stream valleys by clipping their spur ends. Like
streams, tributary and trunk glaciers unite at accordant
levels, but with beds that are discordant about propor-
tionally to the depths of water and ice streams. However,
discordance of beds of glaciers is greatly exaggerated by
comparison with water streams, because of their great
thickness. The surface of a trunk glacier 2000 feet thick
can be joined accordantly by a tributary glacier 500 feet
thick, yet the bed of the side glacier at the point of
junction will be 1500 feet above the bed of the trunk
(Fig 138C).
        Erosion of glaciers, like that of streams, is performed
by the rock material transported. Erosion performed by
glaciers is much more effective than that of streams because of the large size of the material transported, and
the weight of the great thickness of ice above them. The
glacier quarries the rock of its bed, greatest quarrying
of material being in jointed rather than massive rocks.
Quarrying is also greatest in the valley heads, where
cirques are excavated. This is probably accomplished by
alternate freezing and thawing of water on the glacial
floor under great ice pressure. Farther downslope in the
valley glaciers there is both plucking and scouring.
        Shaping of landforms by glaciers is not fully apparent
until the glaciers have largely disappeared from the region
(Figs 138C, 139C).
        Valley-head reservoirs in which the snow and ice had
gathered are enlarged into great excavations, or cirques,
with cliffed sides and headwalls. These encroach from one
or both sides upon the original domed mountain summits;
sometimes a remnant of the original dome remains intact, in
others it is sharpened into a peak (Fig 138C). Encroach-
ment of adjacent cirques may largely consume the divide
between them, leaving it as a saw-toothed ridge, or arête.

A

B

C

Figure 138. Three diagrams illustrating stages of the
   glacial accident: A) Pre-glacial landforms. B) The
   region during glaciation. C) The region after
   glaciation (from Davis, 1906).

Farther down the glaciated valleys, the stream-eroded
slopes are oversteepened and the bottoms become over-
deepened, producing U-shaped valleys. Due to the glacial
habits mentioned above, originally sinuous valleys are
straightened by clipping the spur ends, and tributary
valleys hang above the trunk valley. Valley floors do not
have even gradients, as in stream valleys, but have minor
irregularities, rock falls, sills, and basins. Deepening
of the glaciated valleys commonly decreases toward the
lower ends, where the trough surface may have a gently
ascending slope.

Post-glacial modifications

After disappearance of the glaciers, normal erosional
processes are resumed. Rock basins in the floors of both
cirques and U-shaped valleys are filled by lakes, the
largest of which are commonly behind the ascending slope
at the lower end, where drainage is further dammed by
moraines; these lakes gradually fill up with gravel and
silt. Side streams in the hanging valleys cascade down
into the main valley, and actively cut clefts in a first
attempt to reestablish accordant junctions of branch and
trunk. At the foot of the cascades, alluvial fans are

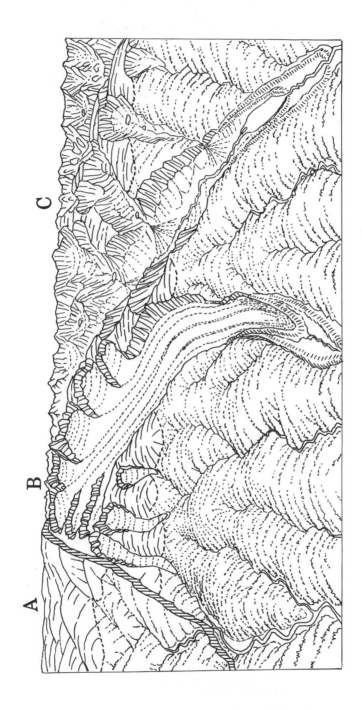

Figure 139. Three-stage diagram illustrating a glacial accident in a region as the Sierra Nevada of California: A) Before glaciation. B) During glaciation. C) After glaciation. Position of timberline during each stage is indicated by heavy dotted line (from Davis, 1933, fig 4).

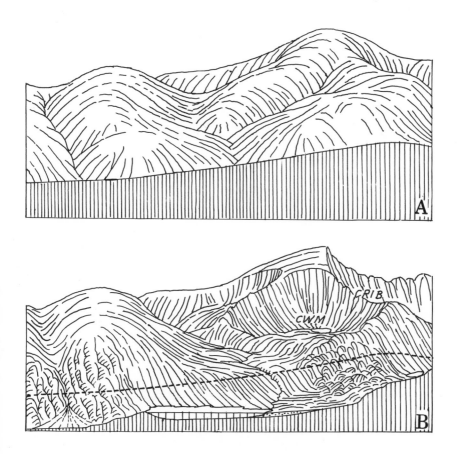

Figure 140. Snowdon, the highest mountain in Wales:
A) The appearance of the mountain before glaciation.
B) Its present appearance, after glaciation (from
Davis, 1912, figs 83, 84).

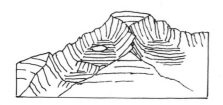
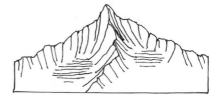

Figure 141. Glaciated peaks in the Front Range of the
Rocky Mountains in Colorado. A) Upland surface
preserved. B) After enlargement of cirques and
destruction of upland remnant (from Davis, 1911,
fig 3).

145

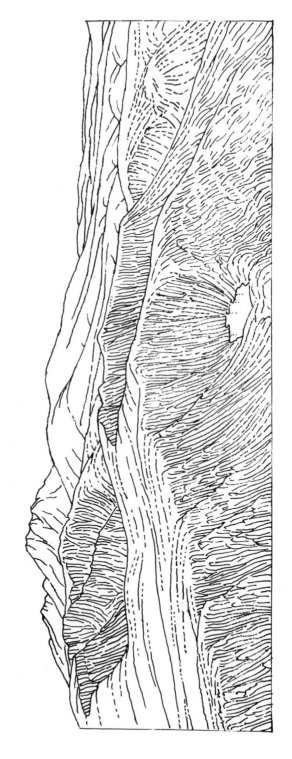

Figure 142. Cirques in the Rocky Mountains in Colorado. View northward along the Continental Divide near Moffat Tunnel northwest of Denver. Arapahoe Peak on skyline to left (from Davis, 1911, fig B2).

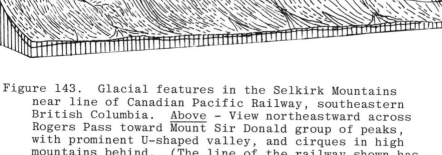

Figure 143. Glacial features in the Selkirk Mountains
near line of Canadian Pacific Railway, southeastern
British Columbia. Above - View northeastward across
Rogers Pass toward Mount Sir Donald group of peaks,
with prominent U-shaped valley, and cirques in high
mountains behind. (The line of the railway shown has
been abandoned since 1913 in favor of the Connaught
Tunnel beneath the Pass). Below - The valley of Beaver
Creek (part of Purcell Trench), on east side of range,
looking west, showing U-shaped valley forms, with minor
valleys meeting the main valley at discordant junctions
(from Davis, 1912, figs 175 and 176).

Figure 144. Glacial forms in the European Alps of France and Switzerland. A) Arêtes and ice fields above the Alpe du Pin in the Vénéon Valley, French Alps (from Davis, 1912, fig 174). B) Hanging valley of the Val Levantina in the lower half of Biasca (from Davis, 1912, fig 167). C) A hanging side valley of the Vénéon Valley, French Alps (from Davis, 1912, fig 168). D) The glaciated Val d'Oisans and the Belladonne Massif (from Davis, 1912, fig 169).

built on the main valley floor, which may push the trunk
streams against the opposing trough wall. The trough sides
are sometimes so oversteepened that great landslides occur,
blocking the trough floors. The oversteepened troughs are
aggraded by streams, especially above the main lakes.

Examples of glaciated regions

Modern mountain glaciers occur in the Alps, Caucasus,
Himalayas, South Island of New Zealand, southern Andes, and
Rocky Mountains (especially in Canada and Alaska). Glacial
forms produced by now-vanished or nearly vanished glaciers
of the preceding glacial period are much more abundant.

Snowdon, the highest mountain in northern Wales (altitude
3560 feet) is a formerly glaciated mountain (Davis, 1909)
(Fig 140). The mountain is formed of deformed massive
rock, which in preglacial times was carved into coarse-
textured, rounded forms of subdued relief. Glaciation
attained maturity. The flanks were carved into great
cirques, locally called cwms (pronounced 'cooms'). Between
some of the cirques are remnants of the preglacial rounded
surface, which now form bridges leading up to the gently
rounded summit. Other cirques have approached one another,
so that the intervening divides are arêtes, locally called
cribs.

The core of the Front Range of the Rocky Mountains in
Colorado is a high level surface of low relief (see Figs
104 and 105), from which project coarse-textured domes of
massive granite. The domes have been maturely glaciated,
and cirques have developed, especially on the leeward or
eastern sides (Figs 141 and 142). In the Rocky Mountains
of Canada, which lie farther north, glaciation has been
more severe, and many valleys have been excavated into
U-shaped forms (Fig 143).

Glacial forms are prominent in the European Alps, a
few samples of the ice field, arêtes, and hanging valleys
are illustrated in Fig 144. The lakes at the south foot of
the Alps, on the boundary between Switzerland and Italy
(Como, Maggiore, Garda, etc.) fill the lower ends of
glaciated valleys (Fig 145). Their bottoms are over-
deepened by glaciers, and their lower southern ends, at
the toes of the former glaciers, are dammed by moraines.
By contrast, the upper surfaces of the glaciers that
deepened the valleys, probably had a continuous downward
slope, clear to their ends. The heads of these lakes,
which extend northward into the Alps, are being filled by
deltas of glacial gravels, washed down from farther back
in the mountains[109].

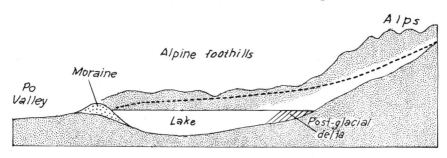

Figure 145. Longitudinal section of a glacial lake at the
south foot of the Alps, on the border of Switzerland
and Italy. Heavy dotted line shows former surface of
the glacier.

NOTES

By Schumm (SAS) and King (PBK)
with assistance from others.

1 The erosion cycle. The cycle of erosion, as advanced
   by Davis, has been attacked and defended by many. For
   reviews of this controversy, see Judson (1960),
   Chorley (1965), Flemal (1971), Higgins (1975), and
   above all, the ultimate evaluation by Chorley et al.
   (1973). Of course, the application of simplified concepts to complex problems will always create controversy and there are many exceptions to Davis's scheme.
   However, the erosion cycle can be observed during
   experimental studies of drainage basin evolution
   (Schumm, 1977), and it obviously can be applied to the
   erosional development of small drainage basins,
   although one may not wish to use Davis's terminology
   when the landforms can be described quantitatively.
       The normal cycle is the progress of the erosional
   evolution of a landscape in humid temperate regions.
   The use of the word 'normal' with regard to this
   environment has evoked reactions of some heat from
   geomorphologists working in drier areas, notably
   L.C. King (1953).   - SAS

2 Instantaneous uplift. Although the concept of
   instantaneous uplift simplifies the model, it was a
   source of considerable criticism. Mountains are not
   formed catastrophically, but nevertheless, if long
   spans of time are considered, the assumption is not
   wholly erroneous because the rate of uplift can be of
   an order of magnitude greater than the rate of denudation (Schumm, 1963).   - SAS

3 Definite and indefinite consequents. These terms have
   not persisted in geomorphologic literature.   - SAS

4 Sink-in. 'Sink-in' means infiltration. This is the
   first of many graphic but unscientific words employed
   by Davis to convey the image of the landscape or an
   understanding of the processes at work. Chorley et
   al. (1973, p 395) comment on his "constant imputation
   of physiological properties in inanimate things". - SAS

5 Drainage texture. For a good example see Carlston
   (1963).

6 Badlands. Drainage density can exceed 400 miles of
   channel per square mile in some badlands. See Smith
   (1957).   - SAS

7 Drainage indefinite and runoff poor. Apparently Davis
   means that runoff will be unconcentrated and subject
   to high infiltration rates.   - SAS

8 Slope morphology. A clear indication, as will be seen
   later (note 67) that Davis accepted the concept of
   control of slope morphology by process. This was
   still being debated in 1956. See Schumm (1956). - SAS

9 Erosion while upheaval is in progress. Davis here is
   reacting to criticism of his assumption of instantaneous uplift.   - SAS

10 Grade. Grade is defined on page 9.

11 If an upheaved lowland stands indefinitely. As pointed
   out by numerous critics this is unlikely; renewed
   tectonism or isostatic adjustment will interrupt the
   cycle.   - SAS

12 Baselevel. Inland river incision cannot approach
   this baselevel, and therefore the peneplain will slope
   toward the sea.   - SAS

13 Peneplain vs. peneplane. D. Johnson (1916) proposed
   to change the spelling of peneplain to 'peneplane' on
   the mistaken assumption that a plain was technically
   composed of flat-lying strata. He and his students
   continued this spelling through the thirties.
   Nevertheless, common usage for 'plain' is any region
   of extensive level ground, and this meaning is
   generally accepted by geomorphologists. In this con-
   text, Davis's definition of peneplain given here and
   elsewhere is clear. The aberrant form 'peneplane' has
   fortunately not survived Johnson.   - PBK

14 Exotic rivers. This, of course, does not apply to
   exotic rivers which flow into progressively drier
   regions, with a decrease in discharge.   - SAS

15 Nearly all the energy of a river is applied to
   carrying its load. Actually only about 5 percent of
   the total energy of a river is used in the transport
   of sediment. See Morisawa (1968, p 411).   - SAS

16 River tends to be broad and shallow. Not necessarily;
   much depends on the nature of the bed and bank sediments and sediment load. In fact, it is more likely
   that a mature channel will have a lower width-depth
   ratio because the sediment load is becoming finer.
   Perhaps Davis means that an alluvial channel is
   broader and shallower than a bedrock channel. - SAS

17 <u>High velocity in upper course of a river</u>. In fact,
mean velocity at mean annual discharge increases in a
downstream direction (Leopold and Madlock, 1953), but
shear stresses on the bed are greater in the steep,
shallow rivers and, therefore coarse sediments are
transported at lower mean velocity.   - SAS

18 <u>Internal and external friction in streams</u>. That is,
roughness of the bed and bank influences velocity as
indicated by the Manning equation. Davis here shows
an intuitive understanding of a basic hydraulic concept, and the following sentence is a qualitative
statement of the continuity equation (see Chow,
1959).   - SAS

19 <u>Obliteration of rapids</u>. Perhaps by burial rather than
entirely by erosion. Even graded streams have steeper
reaches where resistant sediments are introduced
(Hack, 1975, pp 29-58).   - SAS

20 <u>Bend enlargement</u>. The meanders of alluvial streams
are not controlled by variations in the initial course
of the stream, although some incised meanders may be.
Nevertheless, the relation between meander size and
stream size is confirmed by the relations established
between meander wavelength and discharge (Dury,
1964).   - SAS

21 <u>Rise of river during floods, due to excavation of bed</u>.
Local scour can be deep, but in general the average
scour will be only a few centimeters (Colby, 1963).
Deep scour is usually measured at bridges during
floods. Bridges are frequently located at narrow
valley sections, or they cause constriction of flow;
therefore, deeper scour than normal will be recorded
at bridges.   - SAS

22 <u>Rushing rivers</u>. Davis is describing supercritical and
subcritical flow in a steep gradient channel (Chow,
1959).   - SAS

23 <u>Ripple marks in stream bed</u>. Some understandable con-
fusion here concerning bed forms and the hydraulic
conditions under which they form. Standing waves and
antidunes form at supercritical flow. Davis's large
'ripple marks' are dunes (Simons and Senturk, 1977,
chapter 5).   - SAS

24 <u>Capacity of a stream to carry⁄ load varies with the
sixth power of velocity</u>. The size (volume) of the
particles moved, or stream competence is related to
about the sixth power of the velocity (Rubey, 1937),
but sediment load is related to about the third power
of velocity (Colby, 1964).   - SAS

25 <u>Number of threads</u>. Really the surface area of the
water that is in contact with the stream bed. - SAS

26   Busy streams. The busy stream concept makes sense, but Davis appears to be comparing a stream incised into bedrock with an alluvial stream.   - SAS

27   Meander growth. See Figs 17 and 18 for clarification of this statement. The thalweg actually is displaced downstream and meander growth is accompanied by downstream meander shift.   - SAS

28   Vertical contraction of soils. Cohesion of soil particles prevents true vertical contraction; there is always an upslope component of movement during compaction (Kirkby, 1967).   - SAS

29   Rate of creep. This has been confirmed by many quantitative studies (see Carson and Kirkby, 1972; and Young, 1972).   - SAS

30   Recent studies show that much water moves beneath the surface as interflow (Kirkby and Chorley, 1967).
   - SAS

31   Expanding and shrinking rivers. For a discussion of this topic see Gregory and Walling (1973).   - SAS

32   Work of rivers during floods. Major changes in river morphology are accomplished during great floods, but most sediment is transported by numerous floods of moderate size (Wolman and Miller, 1960).   - SAS

33   Carrying power of a young river. An oversimplification; the load depends on the erodability of the bedrock, and sediment loads can be very high as in the steep mountains of New Zealand, South America, and southern California, where shattered rocks provide much sediment.   - SAS

34   Valley-fill deposits. This is a very important but largely ignored concept that explains valley fills at many localities (Schumm, 1977). See discussion by Bloom (1978, p 247). Note also that valley floor aggradation is not shown on Fig 3, although it is shown on earlier Davis figures (Davis, 1899; Schumm, 1977, fig 4-1).   - SAS

35   Underfit rivers caused by aggradation. An interesting and ignored explanation of underfit rivers, which are usually attributed to capture or hydrologic and climatic changes (see Dury, 1964).   - SAS

36   Episodic erosion. Recently it has been suggested that this type of adjustment will be episodic (Schumm, 1977).   - SAS

37   Late stage degradation. The sequence of events described here, as the drainage basin progresses through the erosion cycle from youth (rejuvenation) to old age, has been observed experimentally, (Schumm, 1977).   - SAS

38 <u>Semi-circular meanders</u>. Due to the nature of flow in
river bends most meanders are not semi-circles as
shown in Fig 16B. They have been described as sine-
generated curves (Langbein and Leopold, 1966). - SAS

39 <u>Radius of curvature of meanders becomes larger</u>.
Meanders do not grow in this fashion (Fig 16B); it is
much more likely that they will grow as shown in
Fig 18. See note 27. - SAS

40 <u>River bends become meanders</u>. When a bend becomes a
meander is still a moot question. Leopold and Wolman
(1957) suggest that a sinuosity of 1.5 is required,
but this is too high. - SAS

41 <u>Shift of meanders down-valley</u>. Beautiful examples of
this process are shown on Fisk's Mississippi River
maps (1944), and by experimental studies (Friedkin,
1945). - SAS

42 <u>Amphitheatre</u>. Strictly speaking an amphitheatre is a
double theatre. It is oval or circular rather than
semi-circular. - SAS

43 <u>Slip-off slope</u>. A slip-off slope will not form if
incision is by a stream of high velocity and relatively low sediment load (Shepard and Schumm, 1974).
- SAS

44 <u>Fit river</u>. That is, the valley dimensions are determined by discharge of the stream that formed the
valley. - SAS

45 <u>Effects of cutoff</u>. In fact, the effects of a cutoff
can be considerable, triggering additional cutoffs
both upstream and downstream or causing degradation
upstream and aggradation downstream. - SAS

46 <u>Accretion</u>. Usually termed lateral accretion. - SAS

47 <u>Width of meander belt</u>. Davis is here referring to the
relation between meander amplitude and channel width;
see Leopold, et al. (1964).

48 <u>Deflection of side stream by natural levees</u>.
A Yazoo-type tributary; see page 25. - SAS

49 <u>Underfit rivers</u>. See note 35. - SAS

50 <u>Lengths of winding rivers</u>. The changes have been
documented for the Mississippi River by Fisk (1944)
and recently discussed by Schumm (1977, p 143) who
had forgotten reading the Davis discussion years ago.
A clear case of Gilbert-type plagiarism (Gilbert,
1904). - SAS

51 <u>Graded segment is a reach</u>. 'Reach' as commonly used
designates the length of river under discussion. It
is usually straight. - SAS

52 See note 19.   - SAS

53 Far future, approaching, imminent, etc. An example
of Davis's obsession for descriptive terminology. - SAS

54 Relation of New River. Even more striking are
relations in the Appalachian Valley of southwestern
Virginia, where the highway from Christiansburg to
Radford ascends steeply nearly 600 feet from the head
drainage of the Roanoke River, draining southeastward
into the Atlantic, to a plateau on which flows the
large northwestward-draining New River. The feature
is mentioned briefly by Fenneman (1938, p 261). - PBK

55 Underfitness of Meuse River. Dury's explanation is a
major decrease of discharge produced by climatic
change (Dury, 1964).   - SAS

56 Lateral abstraction. Also termed intercision
(Lobeck, A.K., 1939, p 201).   - SAS

57 Crowley's Ridge. Crowley's Ridge has had a more complex history than implied by Davis. Fisk (1944,
figs 42-54) demonstrates that late-Pleistocene deposition in the Mississippi valley caused the Mississippi
River to spill over the ridge and to join the Ohio
River east of the ridge.   - SAS

58 Desert is any region barren of life. Only by the most
rigorous definition. More commonly it is a dry area
unsuitable for human habitation under natural conditions (AGI Glossary, p 189).   - SAS

59 Lava flows and bare rock surfaces. These are phyto-
logical deserts that exist in areas of high rainfall.
Much of the Earth's surface would have been this type
of desert during pre-vegetation time. - SAS

60 Rainfall is insufficient to fill structural basins.
That is, evaporation, transpiration, and infiltration
exceed precipitation. The definition is restricted
to an area of basin-range topography. - SAS

61 Withering streams. Definition is that for an ephemeral
stream, but Davis also means that there is a decrease
of discharge downstream.   - SAS

62 Alluvial fans. Recognition by Davis of both alluvial
(dry) and fluvial (wet) fans (see Schumm, 1977). - SAS

63 Pediments. Again, clear evidence of Davis's acceptance in 1927 of concepts of parallel slope retreat
and pedimentation (see Davis, 1930). - SAS

64 Integration of basins. Various stages of this process
can be observed in badlands. For example, in Badlands
National Monument pediments coalesce across divides
that are reduced by backwearing of the slopes of
miniature mountain ranges (Smith, 1957). - SAS

65 <u>Deflation basin</u>. The Quattara Depression is a good
example of a major feature of this type (Bloom, 1978,
pp 332-333).   - SAS

66 <u>Conditions seldom if ever realized</u>. Clear recognition
by Davis that his models are based on two assumptions
that are so idealized as to be incorrect. Nevertheless, the assumptions provide a means of dealing with
landscape evolution at a level of sufficient simpli-
city that his model can be developed.   - SAS

67 <u>Parallel slope retreat</u>. Davis recognized parallel
slope retreat 17 years before the symposium that
debated the topic (Von Engeln, 1940, p 222). - SAS
Davis always credited the inception of the idea to
A.C. Lawson (1915).   - PBK

68 <u>Inselberg</u>. Although an inselberg may be composed of
the same material as the surrounding area, the
adjacent area may have more closely spaced joints
(Thomas, 1968).   - SAS

69 <u>Cima Dome</u>. This was selected by A.C. Lawson as his
type 'pan-fan'. The dome is a truly remarkable object
in the landscape, prominently in view from the Union
Pacific Railroad on the south, and from the Los
Angeles-Las Vegas highway on the north. It rises
about 1000 feet above its base and is nearly 10 miles
across, with remarkably smooth slopes, broken only by
a single monticule, Teutonia Peak. But there is much
doubt about its origin, and its validity as a purely
erosional feature is by no means proved. In all
directions around it are rugged, steep-sided mountains
- if it represents the ultimate stage in erosional
degradation, why were not the surrounding mountains
consumed also? Evidence has been cited to suggest
that it may have been formed by tectonic doming
(Sharp, 1957). The other domes farther west in the
Mojave Desert seem more convincing as erosional
features, although there is a possibility of tectonic
warping here also.   - PBK

70 <u>Gravel cover and dissection of pediments</u>. The gravel
cover and the dissection of pediments in southern
Arizona has been variously interpreted by later
observers. In the Ajo region of southwestern Arizona,
Gilluly (1937) interpreted the dissection of pediments
mainly to lowering of gradients during progressive
denudation, and the patchy gravel cover to encroach-
ment from the surrounding bajadas. In the Tucson
area farther east, Tuan (1959) postulated that all the
pediments in that part of the state have been exhumed
from a former gravel cover, but does not explain
clearly the manner in which the underlying rock
surface was formed. - PBK   Recent studies of shale
pediments flanking the Book Cliffs in Colorado and
Utah reveal that the bedrock surface beneath the
gravel is a very irregular stream-eroded surface with
a relief up to 30 feet.   - SAS

71 Prof. A.L. Bloom reminds us that, although this statement is correct, the definitive feature of obsequent streams is that they flow in a direction opposite to the consequent trunk stream.   - SAS

72 Southeastern England. For a complete denudation chronology of this region, see Woolridge and Linton (1955).   - SAS

73 This has been shown by Wills (1938) and Shotton (1953) to be the result of two major glacial diversions.

74 Great Plains. The region is of interest in the history of geomorphology because, Davis tells us, it was here that he came to realize that physical geo- graphy (i.e., geomorphology) had real system and meaning. In 1879, he was recalled to Harvard to take over the teaching of physical geography that had been relinquished by N.S. Shaler. The first year of his new subject (his previous work had been in astronomy and meteorology) proved to be a dreary business, as he was oppressed with the feeling that it was a mere catalogue of unsorted facts ("item-item-item" was his favorite phrase for this kind of business). To add salt to the wounds, President Eliot informed him that the University was little interested in con- tinuing his appointment.

The following summer, however, he was taken on as field geologist for the Tenth Census, under Raphael Pumpelly, and assigned to the Northern Transcontinental Survey, which was investigating the resources along the route of the projected northern transcontinental railroad.

Here, that summer, as outlined above, he grew to realize that these plains were no meaningless surface - merely the waste products of the mountains to the west - but they had had a long prior history and that they had been denuded systematically through time - that initial structures had been subjected to erosional processes, and that erosion had gone through several stages before the present one - that the present plains had been preceded by other plains, now vanished and destroyed.

At the close of the field season, Davis returned to Harvard with a new insight into his subject, determined to find and systematize all of the other landscapes of the Earth.   - PBK

75 Talus below cliff-maker. This may be a bedrock sur- face thinly mantled with rock fragments (Koons, 1955). In addition, many Colorado Plateau sandstones that form high cliffs are, in fact, very susceptible to rapid weathering, as the sand grains are cemented only at grain contacts by calcium carbonate (Schumm and Chorley, 1966).   - SAS

76 Convex outlines of lower cliffs versus concave alcoves in upper cliffs. An excellent example of these relations occurs in the vicinity of King Mountain in western Upton County, west Texas. See Pecos 1:250 000 topographic sheet.   - PBK

77 Truncation of internal structures by range-front faults. Davis (1931, fig 1; reproduced here as Fig 146) presents a view purporting to show truncation of gneissic structure of the southwestern side of the Santa Catalina Mountains by a range-front fault. However, the view was sketched from a hotel rooftop in downtown Tucson, at a distance of 8 miles. Actually, the strike of the gneissic structure is rigidly parallel to the range front, and any apparent truncation is due to perspective and the wish of the beholder. The fault at the base of the range is a thrust, by which the rocks of the lower ground in front have been carried over those of the range. - PBK

Figure 146

78 Persequent streams. This term has not been perpetuated in the AGI Glossary.   - SAS

79 Louderbacks. Davis probably named these features not only in honor of their discoverer, but also because of the 'back' in both 'louderback' and 'backslope'. The story might have been different if his name had been Loudermilk - or Murphy!   - PBK
'Louderback' is unfortunately listed in the AGI Glossary.   - SAS

80 Bight and cusp pattern. A bight and cusp pattern is very common in the range-front faults in the Nevada part of the Great Basin, as shown on the 1:250 000 topographic sheets. Another striking example is along the east face of the fault-block of the Sierra Diablo north of Van Horn, west Texas (King, 1966, p 109 and pl 1).   - PBK

81 Rock walls on eastern side of southern Death Valley. These features are steep curved surfaces which have been called 'turtlebacks' (Curry, 1954). They have been interpreted as the stripped surfaces of overthrust faults, but other interpretations have been made.   - PBK

82 Sequence of forms in arid regions; pan-fans. Compare
   discussion on pages 44-49.   - SAS

83 Age of faulting. If the last stage of Lake Bonneville
   occurred about 10 000 years ago, Davis's equation
   yields a result that is an order of magnitude high.
   For example, Davis (1925) estimated 20 to 200 million
   years were required for the planation of fault-block
   mountains in Utah. Recent estimates of the time in-
   volved in peneplanation involve much shorter periods;
   without isostatic readjustment the continental United
   States would be reduced to base level in about
   10 million years (Gilluly, 1955).   - SAS

84 Fold axes. See Bloom (1978, p 264) for a fuller
   discussion.   - SAS

85 Superimposition of Appalachian drainage. D. Johnson
   (1930) proposed superimposition of Appalachian rivers
   from a former Coastal Plain cover entirely across the
   folded belt, but his proposal has been received with
   little enthusiasm by other geomorphologists. - PBK

86 Erosion cycles in the Appalachian region. It was, in
   fact, in the Appalachians of Pennsylvania that Davis
   (1889) first worked out the principles that form the
   basis of this discussion. The three-stage erosional
   history of the Appalachian region has been widely
   accepted, but various alternative explanations have
   been proposed. Some (for example, Meyerhoff and
   Olmstead, 1936) have proposed many more cycles and
   remnant surfaces. On the other hand, Hack (1960,
   pp 95-96; 1965, pp 4-10) and others have discarded the
   peneplain and cyclical concept entirely, and have proposed that the Appalachian region has progressed in a
   state of equilibrium since the early Mesozoic. This
   does not seem to correspond to the record in the
   adjacent mountain area. However, the validity of the
   alleged Summit peneplain is suspect, because most of
   its remnants are merely the supposed accordant levels
   on narrow ridge crests, which may have been produced
   by other causes. The Valley-floor peneplain, whose
   alleged surfaces are more widely preserved, has more
   to recommend it in my opinion, although estimates of
   its age differ widely.   - PBK

87 Erosion surfaces in the Rocky Mountains. The upland
   of low relief on the summits of the ranges in the
   Southern and Central Rocky Mountains has been called
   the Rocky Mountain peneplain. It is a most impressive
   feature where I have observed it, for example on the
   crest of the Medicine Bow Mountains of southern
   Wyoming. The upland surface (or surfaces) has, however, been variously interpreted. Some observers have
   claimed more than one peneplain level, and van Tuyl
   and Lovering (1935) proposed 8 peneplains, 3 'surfaces'
   (rock benches), and 5 rock terraces in the Front Range
   of Colorado. Another view, widely accepted (Mackin,

1947), is that the upland surfaces are pediments, formed under arid conditions in later Tertiary time, that were originally confluent with the lowland deposits roundabout, and which were cut at altitudes not greatly lower than those at present.   - PBK

88 Colorado piedmont. See Bryan and Ray (1940) for a discussion of the erosional development of the Colorado piedmont.   - SAS

89 Offshore currents. Offshore currents are now considered to operate beyond the nearshore zone. Davis's offshore currents are therefore nearshore currents.   - SAS

90 Wave work. Because of the great variation in wave size, waves can play an important role in sediment transport beyond that indicated by Davis, although the greatest volume of sediment transport by waves takes place at depths of less than 20 meters. However, currents can be generated at depth of 5 kilometers by hurricanes. Velocities of 30 cm/sec were recorded at this depth northeast of Bermuda (EOS, 1977).   - SAS

91 Wave-built terraces. Examples of wave-built terraces are rare if not nonexistent. The continental shelves are areas of sediment transport, not deposition. Wave and current action will prevent development of a depositional terrace (Dietz and Fairbridge, 1968).
                                              - SAS

92 Shorelines of submergence and emergence. The Davis classification was greatly extended and amplified by D. Johnson (1919, 1925), but it is far too simple, as pointed out by many later observers. Davies (1972, p 12) has aptly summarised the later views:
    "... through the first half of the 20th Century the supposed distinction between shorelines of sub- mergence and emergence was widely adopted, but it was abandoned following the realization that post glacial marine transgressions had affected all coasts and only a few are emerged. In fact, shore- lines of emergence are generally a special type of shoreline of submergence which formed where the Holocene seas rose against a gently sloping, relatively undissected landmass."
A better classification would emphasize the degree of equilibrium achieved between coastal morphology and coastal processes.   - SAS

93 Sea cliff and wave cut notch. As sketched this figure would represent rising sea level. A.L. Bloom points out that unless the beach is mantled with talus, the wave-cut notch should remain at sea level.   - SAS

94 Coarse sediments. Usually coarse sediments remain within the surf zone, until abrasion has reduced particle size to about 1 mm. The competency of waves and currents in the nearshore and offshore zones limits the particle size that can be transported to fine sands (T.C. Chamberlain, oral communication, 1977).   - SAS

95 Currents in bays. Currents in bays are very complex. Figure 117 depicts what is considered to be an uncommon circulation pattern.   - SAS

96 Offshore islands tied to shore. These features are tombolos.   - SAS

97 Dry valleys. Modern interpretation is that the now dry valleys contained streams when infiltration was prevented by permafrost. The valleys are relict from Pleistocene conditions (see Gregory and Walling, 1973, p 392).   - SAS

98 Stage. An arbitrary distinction. It is probably not possible for a shoreline to retreat evenly, because of geologic controls and variations in the strength of marine processes.   - SAS

99 Marine planation. Marine abrasion platforms have gradients of about 2 percent, therefore they are not likely to be wider than 1 kilometer (Bradley, 1958).
                                              - SAS

100 400-foot submergence. The result of postglacial sea level rise.   - SAS

101 Sand reef. A barrier, see Bloom (1978, p 457). - SAS

102 Coral reefs. My original notes on this topic were based on one of the first Davis lectures that I heard, given at the University of Iowa in 1925, and they are consequently brief and inadequate. Most of the discussion as given here is, therefore, summarized from Davis's book (1928). No attempt has been made to investigate publications other than those by Davis on this subject since 1928, not to the extensive work that has been done on coral reefs since 1928, and especially those since World War II. It is my impression that most of the new data that have been obtained do much more to confirm the ideas of Darwin, Dana, and Davis, than to confirm those proposed in opposition to them. Menard (1962), in a review of the subject indicates that the recent work largely confirms Darwin's original proposal.   - PBK

103 Holes in lagoons. According to A.L. Bloom they are probably relict Karst.   - SAS

104　Coral reef theories.　One of the instructors at
　　　Harvard, according to Chorley et al. (1973, p 589),
　　　used to diagram the protagonists as follows:
　　　　　　　　　　　DA　RWIN
　　　　　　　　　　　　　NA
　　　　　　　　　　　　　VIS
　　　　　　　　　　　　　LY　　　　- PBK
　　　A.L. Bloom comments as follows, " and these all born
　　　on February 12, if you like such games!"　　- SAS

105　Davis's analysis.　Davis's analysis carries much con-
　　　viction for the simple examples - the volcanic and
　　　coral islands of the tropical oceanic regions - and he
　　　made significant contributions to an understanding of
　　　the marginal belts, as in the West Indies.　His
　　　presentation carries less conviction in the more com-
　　　plex areas, in the tectonically active regions of the
　　　Pacific west of the "andesite line", as in the Fiji
　　　Islands, New Guinea, and Indonesia, where reefs of all
　　　classes abound, but where they and the adjacent lands
　　　have undergone many vicissitudes.　Here, as in his
　　　treatment of other geomorphological problems in
　　　regions that have had a complex history, the usual
　　　Davis oversimplification breaks down; his presentation
　　　becomes a special plea for subsidence and little else,
　　　and details of the geological evidence are brushed
　　　aside.　It is in his discussion of these regions that
　　　the reader loses sympathy with his presentation, and
　　　it is for his discussion of these regions that most
　　　scorn has been heaped on him by later observers. - PBK

106　A.L. Bloom notes that Daly lived to read of the 1954
　　　drilling on Eniwetok and Bikini and to accept sub-
　　　sidence of islands to form atolls.　　- SAS

107　Early history of Rome.　In this geomorphic description
　　　the human elements are quite properly ignored, but
　　　they are a genuine factor.　The description fails to
　　　bring out the original division of the country between
　　　the Latins and the Etruscans.　Only Monte Albano south-
　　　east of the Tiber was in Latin territory, the three
　　　volcanoes to the northwest being in Etruscan territory
　　　- one reason why Davis's story is unfamiliar.　Monte
　　　Albano, with its little side craters of Lake Albano
　　　and Lake Nemi, is deeply rooted in Latin (hence Roman)
　　　tradition.　The significance of Rome - partly cultural,
　　　partly geographical - is that it was the meeting place
　　　of the two nations, and was invigorated by the mixing
　　　of the two cultures.　　　　　　　- PBK

108　Nimble water.　Ice is much more viscous than water with
　　　a viscosity ranging from 10^{10} to 10^{15} poise, depending
　　　on the direction of the applied stress.　Water at $0°C$
　　　has a viscosity of about 18 millipoise.　Hence the
　　　water is millions of times more 'nimble' than ice
　　　(Shumskii, 1964).　　　　　　　　- SAS

109 <u>Conclusion</u>. So, abruptly ends the course. Obviously
     Davis was no better organized than many present-day
     professors who tend to cram several topics into the
     lectures of the last week of the semester.   - SAS

REFERENCES CITED

A.G.I. 1972. *Glossary of Geology*. American Geological Institute, Washington, D.C.

Beckinsale, R.P. 1976. The international influence of William Morris Davis. *Geographical Review,* 66, 448-466.

Blackwelder, E. 1931. Desert plains. *Journal of Geology,* 39, 133-140.

Bloom, A.L. 1978. *Geomorphology*. Prentice-Hall, Englewood Cliffs, NJ.

Bornhardt, W. 1900. *Zur Oberflächengestatung unter Geologie Deutsch Ostafrikas*. Dietrich Reiner, Berlin.

Bradley, W.C. 1958. Submarine abrasion and wave-cut platform. *Geological Society of America, Bulletin,* 69, 967-974.

Bryan, K. 1922. Erosion and sedimentation in the Papago Country, Arizona with a sketch of the geology. *U.S. Geologic Survey Bulletin*, 730-B, 19-90.

Bryan K. and L.L. Ray. 1940. Geologic antiquity of the Lindenmeier site in Colorado. *Smithsonian Miscellaneous Collections*, 99(2).

Carlston, C.W. 1963. Drainage density and streamflow. *U.S. Geologic Survey Professional Paper 422-C*.

Carson, M.A. and M.J. Kirkby. 1977. *Hillslope form and process*. Cambridge University Press, Cambridge.

Chorley, R.J. 1965. A re-evaluation of the geomorphic system of W.M. Davis in Chorley, R.J. and Haggett, P. (eds) *Frontiers in Geographical Teaching*, Methuen, London, 21-38.

Chorley, R.J., Beckisdale, R.P. and A.J. Dunn. 1973. *History of the study of landforms or the development of geomorphology; volume 2, The life and work of William Morris Davis*. Methuen, London.

Chow, V.T. 1959. *Open-channel hydraulics*. McGraw-Hill, New York.

Colby, B.R. 1963. Scour and fill in sand-bed streams. *U.S. Geological Survey Professional Paper 462-D*.

Colby, B.R. 1964. Discharge of sands and mean velocity relationships in sand-bed streams. *U.S. Geological Survey Professional Paper* 462-A, 53-59.

Curry, H.D. 1954. Turtle backs in the central Black Mountains, Death Valley, California. *California Division Mines Bulletin*, 170, 4, 53-59.

Daly, R.A. 1910. Pleistocene glaciation and the coral reef problem. *American Journal of Science*, 4th ser., 30, 297-308.

Daly, R.A. 1915. The glacial-control theory of coral reefs. *American Academy of Arts and Science Proc.*, 51, 155-201.

Daly, R.A. 1919. The coral reef zone during and after the glacial period. *American Journal of Science*, 4th ser., 48, 136-159.

Daly, R.A. 1945. Biographical Memoir of William Morris Davis, 1850-1934. *National Academy of Science Biographical Memoir*, 23, 263-303.

Dana, J.D. 1848. Geology: United States Exploring Expedition during the years 1838, 1839, 1840, 1841, and 1842, under the command of Charles Wilkes, U.S.N., vol. 10.

Dana, J.D. 1853. *On coral reefs and islands*. New York.

Darwin, C. 1942. *The structure and distribution of coral reefs*. London.

Davies, J.L. 1972. *Geographical variations in coastal development*. Oliver & Boyd, Edinburgh.

Davis, W.M. 1889. Rivers and valleys of Pennsylvania. *National Geographic Magazine*, 1, 183-233.

Davis, W.M. 1898. (with W.H. Snyder), *Physical geography*, Ginn & Co., NY.

Davis, W.M. 1899. The Geographical Cycle. *Geographical Journal*, 14, 481-504.

Davis, W.M. 1903. The mountain ranges of the Great Basin. *Harvard University Museum Comparative Zoology, Bulletin*, 42, 129-177.

Davis, W.M. 1904. A flat-topped range in Tian-Shan. *Appalachia*, 10, 277-289.

Davis, W.M. 1906. Sculpture of mountains by glaciers. *Scottish Geographical Magazine*, 22, 76-89.

Davis, W.M. 1908. *Atlas, Practical exercises in physical geography*. Ginn & Co., New York.

Davis, W.M. 1909. Glacial erosion in north Wales. *Geological Scoiety of London Quarterly Journal*, 65, 281-350.

Davis, W.M. 1911. The Colorado Front Ranges. *Association of American Geographers Annals*, 1, 21-84.

Davis, W.M. 1912. *Die erklärende Beschriebung der Land-formen*. B.G. Teubner, Berlin.

Davis, W.M. 1925. The Basin Range problem. *U.S. National Academy of Science Proceedings*, 11, 387-392.

Davis, W.M. 1926. The Lesser Antilles. *American Geographical Society Map of Hispanic America, Publication* 2.

References

Davis, W.M. 1928. The coral reef problem. *American Geographical Society special publication* 9.

Davis, W.M. 1930a. Rock floors in arid and humid climates. *Journal of Geology*, 38, 1-27, 136-158.

Davis, W.M. 1930b. The Peacock Range, Arizona. *Geological Society of America, Bulletin* 41, 293-313.

Davis, W.M. 1930c. Physiographic controls, east and west. *Science Monthly*, 30, 395-415, 501-519.

Davis, W.M. 1931. The Santa Catalina Mountains, Arizona. *American Journal of Science*, 5th series, 22, 289-317.

Davis, W.M. 1933. Granitic domes of the Mojave Desert, California. *San Diego Society of Natural History, Transactions* 7, 211-258.

Davis, W.M. 1938. Sheet floods and stream floods. *Geological Society of America, Bulletin* 49, 1337-1416.

Davis, W.M. and B. Brooks. 1930. The Galiuro Mountains, Arizona. *American Journal of Science*, 5th series, 19, 89-115.

Dietz, R.S. and R.W. Fairbridge. 1968. Wave base, in: *Encyclopedia of geomorphology*. Reinhold Book Corp., New York, 1224-1228.

Dury, G.H. 1964. Principles of underfit streams. *U.S. Geological Survey Professional Paper* 452-A.

EOS. 1977. *American Geophysical Union Transactions*, 58, 956.

Fenneman, N.M. 1938. *Physiography of eastern United States*. McGraw-Hill Book Co., New York.

Fisk, H.N. 1944. *Geological investigation of the alluvial valley of the lower Mississippi River*. Mississippi River Commission, Vicksburg, Mississippi.

Flemal, R.C. 1971. The attack on the Davisian system of geomorphology, a symposium. *Journal of Geological Education*, 19, 1-13.

Friedken, J.F. 1945. *A laboratory study of the meandering of alluvial streams*. U.S. Army Waterways Experiment Station, Vicksburg, Mississippi.

Gilbert, G.K. 1875. Report on the geology of portions of Nevada, California, and Arizona (1871-1872), in: Report on Geographical and Geological Explorations and Surveys West of the One Hundredth Meridian (Wheeler Survey). *Geological Society of America, Bulletin* 48(3), 323-348.

Gilbert, G.K. 1890. Lake Bonneville. *U.S. Geological Survey Mon.* 1.

Gilbert, G.K. 1904. A case of plagiarism. *Science*, n.s. 20, 115-116.

Gilluly, J. 1937. Physiography of the Ajo region, Arizona. *Geological Society of America, Bulletin* 48(3), 323-348.

Gilluly, J. 1955. Geologic contrasts between continents and ocean basins. *Geological Society of America Special Paper* 62, 7-18.

Gregory, K.J. and D.E. Walling. 1973. *Drainage basin form and process*. John Wiley, New York.

Gulliver, F.P. 1896. Cuspate forelands. *Geological Society of America, Bulletin* 7, 349-422.

Hack, J.T. 1965. Geomorphology of the Shenandoah Valley, Virginia and West Virginia. *U.S. Geological Survey Professional Paper* 484.

Higgins, C.G. 1975. Theories of landform development - a perspective, in: Melhorn, W.N. and R.C. Flemel (eds) *Theories of landform development*. Publications in Geomorphology, State University of New York, Binghampton, NY, 1-28.

Johnson, D.W. 1916. Plains, planes, and peneplains. *Geographical Revue*, 1, 443-447.

Johnson, D.W. 1919. *Shore processes and shoreline development*. John Wiley, New York.

Johnson, D.W. 1925. *The New England-Canadian shoreline*. John Wiley, New York.

Johnson, D.W. 1931. *Stream sculpture on the Atlantic slope; a study in the evolution of Appalachian rivers*. Columbia University Press, New York.

Johnson, W.D. 1901. The High Plains and their utilization. *Geological Survey 21st Annual Report*, part 4, 609-741.

Judson, S. 1960. William Morris Davis; an appraisal. *Zeit. Geomorph.* 4, 193-201.

Keyes, C.R. 1909. Erosional origin of the Great Basin Ranges. *Journal of Geology*, 17, 31-37.

Keyes, C.R. 1912. Deflative scheme of the geographic cycle in an arid climate. *Geological Society of America, Bulletin* 23, 537-562.

King, L.C. 1953. Canons of landscape evolution. *Geological Society of America, Bulletin* 64, 721-732.

King, P.B. 1966. The geology of the Sierra Diablo region, Texas. *U.S. Geological Survey Professional Paper* 480.

Kirkby, M.J. 1967. Measurement and theory of soil creep. *Journal of Geology*, 75, 359-378.

Kirkby, M.J and R.J. Chorley. 1967. Throughflow, overland flow and erosion. *International Association of Science Hydrology Bulletin*, 12, 5-21.

References

Koons, D. 1955. Cliff retreat in southwestern United States. *American Journal of Science*, 253, 44-52.

Langbein, W.B. and L.B. Leopold. 1966. River meanders; theory of minimum variance. *U.S. Geological Survey Professional Paper* 422-H.

Lawson, A.C. 1915. The epigene profiles of the desert. *California University Department of Geology Bulletin*, 9, 23-48.

Leopold, L.B. and M.G. Wolman. 1957. River channel patterns; braided, meandering, and straight. *U.S. Geological Survey Professional Paper* 282-B.

Leopold, L.B. and T. Maddock. 1963. Hydraulic geometry of stream channels and some physiographic implications. *U.S. Geological Survey Professional Paper* 352.

Leopold, L.B., Wolman, M.G. and J.P. Miller. 1964. *Fluvial processes in geomorphology*. W.H. Freeman, San Francisco, California.

Lobeck, A.K. 1939. *Geomorphology*. McGraw-Hill, New York.

Louderback, G.D. 1904. Basin range structure in the Humboldt region. *Geological Society of America Bulletin*, 15, 289-346.

Mackin, J.H. 1947. Altitude and local relief in the Bighorn area during the Cenozoic, in: *Field conference in Bighorn Basin guidebook*. Wyoming University, Wyoming Geological Association, and Yellowstone-Bighorn Research Association, 103-120.

McGee, W.J. 1897. Sheetflood erosion. *Geological Society of America, Bulletin* 8, 87-112.

Menard, H.W. 1962. Foreword, in: Darwin, C., *The structure and distribution of coral reefs* (reprint). California University Press, Berkeley and Los Angeles, v-ix.

Meyerhoff, H.A. and E.W. Olmstead. 1936. Origins of Appalachian drainage. *American Journal of Science*, 5th series, 32, 21-42.

Morisawa, M. 1968. *Streams, their dynamics and morphology*. McGraw-Hill, New York.

Murray, J. 1880. On the structure and origin of coral reefs and islands. *Royal Society of Edinburgh Proceedings*, 10, 505-518.

Murray, J. 1887-1888. Structure and origin of coral reefs and islands. *Royal Institute Proceedings*, 12, 251-262.

Paige, S. 1912. Rock-cut surfaces in the desert region. *Journal of Geology*, 20, 442-450.

Passarge, S. 1904. *Die Kalahari*. Berlin.

Powell, J.W. 1875. *Exploration of the Colorado River of the West*. Washington, DC.

Rubey, W.W. 1937. The force required to move sediments in a stream bed. *U.S. Geological Survey Professional Paper* 189-E, 121-143.

Russell, I.C. 1884. A geological reconnaissance in southern Oregon. *U.S. Geological Survey 4th Annual Report*, 431-464.

Schumm, S.A. 1956. The role of creep and rainwash on the retreat of badland slopes. *American Journal of Science*, 254, 693-706.

Schumm, S.A. 1963. Disparity between present rates of denudation and orogeny. *U.S. Geological Survey Professional Paper* 454-H.

Schumm, S.A. 1977. *The fluvial system*. Wiley-Interscience, New York.

Schumm and R.J. Chorley. 1966. Talus weathering and scarp recession in the Colorado Plateau: *Zeit Geomorphologie*, 10, 11-36.

Sharp, R.P. 1957. Geomorphology of Cima Dome, Mojave Desert, California. *Geological Society of America, Bulletin* 68, 273-289.

Shepard, R.G. and S.A. Schumm. 1974. Experimental study of river incision. *Geological Society of America, Bulletin* 85, 257-268.

Shumskii, P.A. 1964. *Principles of Structural Glaciology*. Dover, New York.

Simons, D.B. and F. Senturk. 1977. *Sediment transport technology*. Water Resources Publications, Fort Collins, Colorado.

Smith, K.G. 1957. Erosional processes and landforms in Badlands National Monument, South Dakota. *Geological Society of America, Bulletin* 69, 975-1008.

Spurr, J.E. Origin and structure of the Basin Ranges. *Geological Society of America, Bulletin* 12, 217-270.

Thomas, M.F. 1968. Bornhardts, in: *Encyclopedia of Geomorphology*. Reinhold Book Corp., New York, 88-90.

Tolman, C.F. 1909. Erosion and deposition in the southern Arizona bolson region. *Journal of Geology*, 17, 136-163.

Tuan, Y.F. 1959. Pediments of southeastern Arizona. *California University Publications in Geography*, 13.

Von Engeln, O.D. 1940. Symposium; Walter Penck's contribution to geomorphology. *Association of American Geographers Annals*, 30, 219-284.

Van Tuyl, F.M. and T.S. Lovering. 1935. Physiographic development of the Front Range. *Geological Society of America, Bulletin* 46, 1291-1350.

References

Wolman, M.G. and J.P. Miller. 1960. Magnitude and frequency of forces in geomorphic processes. *Journal of Geology*, 68, 54–74.

Woolridge, S.A. and D.L. Linton. 1955. *Structure, surface, and drainage in southeast England*. George Philip, London.

Young, A. 1972. *Slopes*. Oliver & Boyd, Edinburgh.

INDEX

APPENDIX

NATIONAL ACADEMY OF SCIENCES
OF THE UNITED STATES OF AMERICA
BIOGRAPHICAL MEMOIRS
VOLUME XXIII—ELEVENTH MEMOIR

BIOGRAPHICAL MEMOIR

OF

WILLIAM MORRIS DAVIS
1850–1934

BY

REGINALD A. DALY

PRESENTED TO THE ACADEMY AT THE AUTUMN MEETING, 1944

WILLIAM MORRIS DAVIS

1850–1934

BY REGINALD A. DALY

William Morris Davis won distinction as geologist, meteorologist, and geomorphologist, but primarily as teacher. He made personal, outdoor researches in every continent except Antarctica, as well as in island groups of the Atlantic and the Pacific; yet his international fame rests chiefly on his development of a system of thought concerning the reliefs, the scenery, of our planet. His system is the "American" system, but it is applicable to the landscapes of the whole world. His early training in geology led him to the principle by which he, more than anyone else, has revolutionized the teaching of, and research on, the endlessly varied forms of the lands and coastlines. To geographers and geologists alike he was an apostle bringing to them the gospel of method in research and method in the presentation of the results of research. For him the root of the matter is evolution, orderly development. Many geologists had used this principle, so essential to understanding the protean crust of the earth, but few geographers had used it in describing land forms. Davis emphasized a mode of thinking and for its expression he devised a system which has greatly appealed to teachers and investigators in many foreign countries as well as in the United States of America. While creating his descriptive method in terms of evolutionary changes, he found our English tongue sadly deficient. He had to create a new, necessarily technical language. Every man of science knows the difficulty of such an invention. Some of his verbal tools Davis was able to adopt from the literature of earth science, an immense literature which he thoroughly mastered; other vital terms were his own. The combination has been put to constructive use by geologists and geographers, foreign and domestic, to an extent encouraging to our pioneer. He lived to see notable improvement of geographical instruction in grammar school, high school, college, and university; improvement in the reporting of geographical and geological facts by staffs of the State and Federal surveys; and improvement in the discussion of "terranes" by the more philosophically-minded historians and economists.

Knowing that even a long life could not vitalize all the dry bones of the old geography, Davis specialized on physical geography, leaving to others the problem of systematizing the infinitely varied responses of organisms to their environment. This other half of geography needs today a clarifying leader like Davis.

Davis was born in Philadelphia on February 12, 1850. At that time his father, Edward M. Davis, a business man, and his mother, Maria (Mott) Davis, were members of the Society of Friends and fully shared the best characteristics and activities of Quakers in full standing. Yet a hatred of injustice, which was to be an outstanding emotion of their son, led them into rebellion against one of the ironclad rules of the Society. Not content with helping to operate the "underground railway" for escaping slaves, Edward Mott enlisted in the Northern Army. For this action he was expelled from the Society of Friends and soon after his wife resigned from it. To both of them the question of States' rights was quite subordinate to the problem of human freedom. To break with the Society's tradition took courage of the kind shown in the remarkable life of Lucretia Mott, the mother of Maria Mott.

Theodore Tilton called Lucretia Mott, born in Nantucket, Massachusetts, "the greatest woman ever produced in this country." "She was the real founder and the soul of the woman's rights movement in America and England. She was the outstanding feminine worker in the struggle to rid our country of slavery. She advocated labor unions in a day when they were proscribed and generally considered illegal. She proscribed war, and worked diligently for liberal religion." Her crusading force "had its source in the love of freedom of her seafaring ancestry, and she feared opposition or the exploration of uncharted regions of the mind no more than they feared to venture into unknown seas" (quotations from Anita Moffett in the New York Times Book Review, August 1, 1937). That her grandson was to be a crusader, a champion for mental and moral probity, was assured by inheritance from parents and grandparent.

As a boy Davis was retiring, little interested in sports, but engrossed in his studies. For several years before attending the local schools he was taught his lessons by his mother. She, like her own mother, knew well the power of words and laid much stress on their correct use; doubtless this early training had much to do with Davis's rigor in developing a scientific vocabulary for his favorite science and his insistence on precision of speech and writing by student or professional investigator.

The boy was a good student and showed his mental calibre by winning the Harvard degrees of Bachelor of Science at the age of nineteen and Master of Engineering a year later. He immediately accepted a call to the meteorological service of the National Observatory of Argentina at Cordoba. After three years of that routine work he returned to the United States. After a term as field assistant to Pumpelly in the Northern Pacific Survey, he was appointed (1877) to an assistantship in geology at Harvard, under N. S. Shaler, with whom he gained a permanent love for earth science. In those days promotion was slow and from 1879 to 1885 he was listed as instructor in geology at Harvard, where he began a five-year term as assistant professor of physical geography in 1885. In 1890 he attained the rank of full professor in the same subject. Nine years later he became Sturgis Hooper Professor of geology, a position held until 1912, when he resigned, to be a Harvard "emeritus" for the remaining twenty-two years of his life. He had two leaves of absence. In 1908 he was appointed visiting professor at Berlin University for a year, and, in 1911, visiting professor at Paris for a year, during which he lectured also at several provincial universities of France.

With his resignation Davis was freed from his responsibilities as active Sturgis Hooper Professor and found the eagerly-sought opportunity to make many postponed field studies both in North America and abroad, and also to make personal contact with geographers and geologists and their respective workshops. Because he had a philosophy to expound, he could not refrain from accepting many invitations to lecture at western universities: California (Berkeley, 1927-1930); Arizona (1927-1931); Stanford (1927-1932); Oregon (1930); California Institute of Technology (1931-1932). With unfinished manuscripts on his desk at Pasadena he died in harness, on February 5, 1934, seven days before his eighty-fourth birthday.

The efficiency of Davis as a man of science was in no small part secured by domestic happiness. He was married three times and twice he suffered by the death of a partner. In 1879 he married Ellen B. Warner of Springfield, Massachusetts; in 1914, Mary M. Wyman of Cambridge, Massachusetts; and in 1928, Lucy L. Tennant of Milton, Massachusetts, who has survived him. All three women were truly sympathetic helpmeets, as the present writer knows from forty years of close association with this man, who needed much freedom from the cares of a household while working for and in the world outside.

The science of land forms, so intelligently enriched and organized by Davis is a planetary science; his message was addressed to geographers and geologists of every nation. That those colleagues recognized the vitality and soundness of his evolutionary ideas is indicated by the long list of honors showered on him by foreign as well as American societies. He was elected to honorary membership in the geographical societies of Amsterdam, Berlin, Budapest, Frankfurt, Geneva, Greifswald, Leipzig, Madrid, Neuchatel, New York, Petrograd, Rome, Stockholm, and Vienna, as well as the Royal Society of Natural History at Madrid, the American Meteorological Society, and the Scientific Society "Antonio Alzate" of Mexico; to corresponding-membership in the Berlin and Paris Academies of Science and the Accademia dei Lincei; to corresponding-membership in the geographical societies of Chicago, London, Munich, Paris, and Philadelphia, and the geological societies of Belgium, Liverpool, and London and the German Meteorological Society; to foreign-membership in the Academies of Sciences at Christiania, Copenhagen, and Stockholm. He was elected to membership in the American Academy of Arts and Sciences, the American Philosophical Society, the National Academy of Sciences, the Imperial Society of Natural History in Moscow, and the New Zealand Institute.

The Geological Society of America made him its acting president in 1906 and full-time president in 1911. He founded the Harvard Travelers Club, of which he was president from 1902 to 1911, and the Association of American Geographers, of which he was thrice elected president (1904, 1905, 1909). For his leadership and scholarliness he was chosen to be an associate editor of "Science" and the "American Journal of Science."

In 1886 he was awarded the H. H. Warner Medal "for scientific discovery"; in 1895, another from the University of Paris. Later awards were: the Cullom Medal from the American Geographical Society (1908); a medal from the University of Berlin (1910); a medal from the Harvard Travelers Club (1911); a medal from the Geographical Society of Philadelphia (1912); the Culver Medal from the Geographical Society of Chicago (1913); the Kane Medal from the Philadelphia Geographical Society (1913); the Hayden Medal from the Philadelphia Academy of Sciences (1918); the Patron's Medal from the Royal Geographical Society, London (1919); the Vega Medal from the Swedish Geographical Society (1920); the Loçy Jagos Medal from the Hungarian Geographical Society (1930); and the Penrose Medal from the Geological Society of America (1931). He became Chevalier of the French Legion of Honor. As Exchange Professor to France he was the first American to give regular instruction at the Sorbonne.

Davis was given four honorary degrees: S.D. by the University of the Cape of Good Hope (1905) and by the University of Melbourne (1914); Ph.D. by the universities of Greifswald (1906) and Christiania (1911).

After his death the California Institute of Technology at Pasadena, where he had made many new friends, dedicated to Davis a memorial "Gate of Knowledge," one of the entries to the grounds of the Institute, whose students and faculty he had stimulated by his courses of lectures.

Work in Meteorology

Davis's interest in meteorology was doubtless aroused by his study of atmospheric conditions in Argentina, from 1870 to 1873. Soon after his appointment at Harvard he undertook his first pioneering task, the creation of a systematic course on the science of the atmosphere. This course became noted for its broad scope and for the clear, logical mode of presentation; in these respects it had no rival in America and probably none anywhere else. Fortunately he was able to put the content of the course in the permanent form of his "Elementary Meteorology," published in 1894, when the course was turned over to Robert DeCourcy Ward, a capable, Davis-trained student, who greatly expanded the university offerings in meteorology and added courses in climatology. This development, together with the founding of the Blue Hill Observatory as a Harvard research institution, was an abiding satisfaction to Davis and incidentally freed him for other enterprises.

The superbly designed and executed "Elementary Meteorology," for many years the best college text on the subject and still valuable in spite of the enormous increase of meteorological data since 1894, illustrated its author's skill in compiling the best of the world's thought about the physics of the atmosphere and contained the results of his own direct observations. With the help of volunteer assistants he carried on such field investigations as could be prosecuted in New England. The results were published in papers on thunderstorms, the sea breeze, atmospheric convection, and theories of rainfall. Other papers with novel points of view were published on tornadoes, secular changes of climate, and the wind systems of the oceans. His writings on thunderstorms and the sea breeze are "classic" for teachers of meteorology. Between 1884 and 1893 he published forty papers on this general subject.

Work in Geology

Not long after Davis became associated with the inspiring Shaler, the two men published jointly a handsome volume "Illustrations of the Earth's Surface" (1881), intended to popularize some of the more dramatic and better understood processes that mold the surface of our planet. But the young instructor knew full well that effective, authoritative teaching of geology, the principal subject of his first instructorship, demanded close personal touch with Nature. To get such experience he selected for field study in detail the Triassic formation of New England and New Jersey. On those regions he published fifteen preliminary papers (1882-1896), and a monographic summary of most of his results in "The Triassic Formation of Connecticut" (1898). This gave the first full account of the Triassic volcanic history of the region, announced criteria for proving the extrusive character of some of the "trap sheets" and the intrusive character of others. He also showed how the analysis of topographic forms could be used in explaining the underground, invisible structures of Connecticut and similarly faulted areas of the earth's crust.

While working on the complex history of the Triassic areas, Davis interpolated field investigations: in Columbia County, New York, and the Catskills, where he described the northward continuation of the Appalachian structure; on the glacially-formed drumlins of New England and other regions; on the structure and origin of glacial sandplains and eskers; and on the geological history of Mount Desert Island. In later years he studied: the origin of the thick and widespread Tertiary formations of the Rocky Mountain region, showing that these are not lake beds, as had been generally assumed, but are fluviatile and alluvial-fan deposits; the origin and erosional history of the Basin Ranges of the West; the development of the Colorado Canyon; the mechanical conditions leading to the formation of limestone caverns; and the nature of geological proof, asking geologists "how do you know you are right?"— a question that illustrated the fact that he was as much concerned with the method of scientific thinking as he was in the majestic happenings of earth history. Yet Davis must have been conscious that he made a principal contribution to the philosophy of geology itself. His major contribution to earth science was the conception of the "erosion cycle." He applied it to the physiographic history of Pennsylvania, New England, the Rhine province, Turkestan, and many other, once-lofty ranges of mountains and proved that each of these regions had been reduced by slow denudation to a lowland, to an "almost-plain" or "peneplain." He further showed that after completion of a cycle, many an "old-mountain peneplain" was uplifted and again deeply dissected by its rivers. With such demonstrations, phrased in the terms of his new geographical vocabulary, Davis made more vivid than ever before the enormous length of geological time. No geologist who had carried the logic of the erosion cycle into the interpretation of the major "unconformities" visible in the strata of the earth's crust was greatly surprised when, later, the results of radioactivity in rocks gave a minimum age of about two billion years to that crust.

Work in Geography

Davis gave much thought to the question as to the content of scientific geography, a subject which, because of the world-wide problems of both war and peace, is likely to be in long-continued demand in our colleges and universities as well as in secondary schools. In the first yearbook of the National Society for the Scientific Study of Education (1902) he wrote:

"Geography as a mature subject is capable of a higher development than it has yet reached. In this connection it will be well to review briefly the three stages of development recognizable in the progress of our venerable subject. Until within about a hundred years the content of geography consisted of a body of uncorrelated facts concerning the earth and its inhabitants. The facts were described empirically, and as a rule very imperfectly. Their location was noted, but their correlations were overlooked; it had not indeed been clearly made out that correlations existed. This blindly inductive first stage was followed by a second stage, which was opened by Ritter's exposition between the earth and its inhabitants; . . . such relationships as were noted had to be explained on the old doctrine of teleology—the adaptation of the earth to man—instead of on the modern principle of evolution—the adaptation of all the earth's inhabitants to the earth. It is this principle which characterizes the third stage of progress, and along with it goes a principle of almost equal importance; namely, that all the items which enter into the relation between the earth and its inhabitants aid so powerfully in observing and appreciating the facts of nature." . . .

"Geography has today entered well upon its third stage of progress. The 'causal notion' is generally admitted to be essential in the study. . . . Thus understood, geography involves the knowledge of two great classes of facts: first, all those facts of inorganic environment which enter into relationship with the earth's inhabitants; second, all those responses by which the inhabitants, from the lowest to the highest, have adjusted themselves to their environment. The first of these classes has long been studied as physical geography, although this name has been used as a cover for many irrelevant topics. In recent years there has been a tendency to compress the name into the single word 'physiography.'

"The second of the two classes of facts has not yet reached the point of being named, but perhaps it may come to be called ontography. Ecology, to which increasing attention is given by biologists, is closely related to what I here call ontography, yet there is a distinction between the two, in that ecology is concerned largely with the individual organism, while ontography is intended to include all pertinent facts in structure, physiology, individual, and species.

"Neither physiography nor ontography alone is geography proper, for geography involves the relation in which the elements of its two components stand to each other. Each of the components must be well developed before geography can be taken up as a mature study."

Davis held that "teachers of geography should be better taught"; that the subject should be treated more scientifically both here and abroad; that it is far more than the "location of things"; that emphasis on principles rather than on items cannot fail to foster the "intelligence as well as the memory" of pupils in secondary schools; that even in such schools the causal notion should be stressed—"how" and "why" as well as "where" and "what," about things as we find them. "Elementary geography may still deal with the salient facts and place man conspicuously in the foreground; more advanced geography may include examples of greater complexity, by always selecting important rather than trivial matters; but the investigator must study the trivial items along with the greater ones, and all must be duly scrutinized, described, and classified."

The delay of the subject to reach mature treatment did not surprise Davis, who regarded it as "perhaps the most complex of all sciences." Although he did not mention it, not the least of the complications in human geography is man's free will, so often obscuring his responses to physiographic controls. Thus for more than one reason Davis himself did comparatively little in illustrating his fundamental principle of relationship between organisms and environment. He wisely restricted himself to spade work on the inorganic side of the vast subject.

In his chosen field Davis worked on the principle that, while geology is the study of the past in the light of the present, physiography is the study of the present in the light of the past. The one science complements the other and it is no accident that his influence on geological research has been at least as great as his influence on geographical research.

On many occasions he told of his deeply-felt indebtedness to American geologists, particularly Lesley, the staff of the Geological Survey of Pennsylvania, and Powell, Gilbert, Dutton and Holmes of the great western surveys. It was while reading their published writings that "geography gained a new interest" for Davis. That interest culminated in the development of his most famous idea, that of the "cycle of erosion." He visualized a structural unit in the terrestrial landscape and then deduced the topographic results of erosion of this unit by rivers born on its original surface or developed on the unit during the later, systematic evolution of its river system.

"The sequence of forms assumed by a given structure during its long life of waste is determinate, and . . . the early or young forms are recognizably different from the mature forms and the old forms. A young plain is smooth. The same region at a later date will be roughened by the channeling of its larger streams and by the increase in number of side branches, until it comes to 'maturity,' that is to the greatest variety or differentiation of form. At a still later date the widening of the valleys consumes the intervening hills, and the form becomes tamer, until in 'old age' it returns to the simple plain surface of 'youth' " (*National Geographic Magazine,* vol. 1, 1888, p. 15).

In another place he wrote:

"In whatever way a new mass is offered to the wasting forces, let us call the forces that uplift it constructional forces; and the forms thus given, constructional forms. Let all the forces of wasting be called destructional forces; let the sea-level surface, down to which a sufficiently long attack of the destructional forces will reduce any constructional form, be called the ultimate baselevel; and let the portion of geological time required for the accomplishment of this task be called a geographical cycle. Construction, destruction, baselevel and cycle are our primary terms." (*Journal of Geology*, vol. 2, 1894, p. 72.)

It should be noted that "cycle" is here used in the figurative sense of a long period of time. The "plain" of extreme old age could never attain the form of the youthful stage, the greatly multiplied branches of the master rivers and also the interstream areas having individual slopes quite different from the general slope of the young plain, both in magnitude and azimuth. Thus at the ultimate stage of development of the ideal cycle we have an almost-plain with a relief which, though gentle, is vastly more varied than the relief of the young plain. To this final form Davis gave the name "peneplain," which, like "cycle," has won a permanent place in the vocabulary of physiographers and geologists.

Similarly, Davis worked out the ideal cycle as a means of vividly describing the erosional changes suffered by terranes of much greater variety of initial relief, such as mountain ranges and volcanic provinces. With sufficient study any actual unit of the earth's topography can be interpreted in terms of the erosion cycle, with its three dominating ideas, structure, process, and stage.

Nevertheless Davis knew well that the scheme of a simple cycle can rarely suffice for a full scientific description of land forms. He saw that at any stage of its history a topographic unit may be affected by uplift or subsidence, with corresponding effect on the power of eroding streams and on the fashioning of reliefs. Thus the deductive scheme was enlarged to the conception of multiple cycles, separated by "interruptions" due to changes of level. Then, too, the landscape in question may have had its drainage system affected by change of climate or by volcanism—complications to which he gave the technical name "accidents."

In the Proceedings of the American Philosophical Society (1902) Davis further explained his mode of thought as teacher and investigator in the following words:

The geographer "must generalize in order to bring the observable items within the reach of descriptive terms, and as soon as he generalizes, the use of idealized types is practically unavoidable. Such types have long been in current use, but they have been too few and too empirically defined for the best results. They need to be greatly increased in number, and at the same time they must be correlated with structure, process, and time; for only by following the path of nature's progress can we hope to store our minds with types that shall imitate nature's products. It may be fairly urged that the larger the store of types a geographer possesses, and the more careful and numerous the comparisons with nature by which the types have been rectified, the better progress can the geographer make in new fields of observation.

"But the geographer who adopts the explanatory methods in a whole-souled fashion will find himself called upon not only to imagine a large series of type forms; he must also call into exercise his deductive faculties and employ them to the fullest, if he would make the best progress in the newer phases of his subject, however purely inductive he has imagined it to be. In setting up a store of types, there is need of deducing one type from another at every step; and it may be confidently urged that whoever hesitates to recognize this principle will fail of his effort to describe through explanation. But as a matter of fact, geography has some time been more deductive than geographers have supposed it to be; and the newer phase of the science is not characterized so much by introducing deduction for the first time, as by insisting on its whole-souled acceptance as an essential process in geographical research.

"It is only by giving the fullest exercise to the faculties of imagination and deduction that the cycle of erosion becomes serviceable. Here the geographer who hesitates is lost. . . .

"Thus comparing the partial view of the landscape, as seen by the outer sight, with the complete view of the type as seen by his inner sight, [the geographer] determines, with great saving of time and effort, just where his next observations should be made in order to decide whether the ideal type he has provisionally selected fully agrees with the actual landscape before him. When the proper type is thus selected, the observed landscape is concisely and effectively named in accordance with it; and description is thus greatly abbreviated."

As he put the case in 1894, "one of the chief aids to sharp outsight is clear insight." To illustrate, he cited the need of special training for the maker of topographic maps.

"Even the best surveys are necessarily sketched in great part; and the topographer must appreciate his subject before he can sketch it. He must have a clear insight into its expression; his outer eye must be supplemented by his inner eye." . . . Let us therefore strive to complete a deductive geographical scheme . . . until it shall at last be ready to meet not only the actual variety of nature, but all the possible variety of nature."

Davis gave still another summary of the method he recommended to the geographer who aspired to be truly scientific. The savage may do little more than observe natural happenings. The barbarian may go a step further and invent hypotheses in explanation of those events; although his hypotheses are generally wild, he may be said to have a two-faculty approach to Nature. The modern, well-trained naturalist takes four steps. He observes, invents, deduces, and verifies; he deduces the consequences of each hypothesis and then goes back to Nature to improve his deductive scheme and to verify the correct hypothesis if he has been fortunate enough to create it. He has the four-faculty approach to Nature. Two generations of workers in earth science have benefited by Davis's insistence on the value of multiple hypotheses, even "outrageous" hypotheses, in search for the truth about the outdoor world. By such thinking all around the subject, that is, by inventing all of the more reasonable, conceivable solutions to the problem at issue, the investigator is put on the alert. His field record becomes automatically charged with crucial observations and kept free from a load of hit-or-miss, unessential observations. Valuable as it is, the scheme of the erosion cycle is not so important for research in earth science as the underlying philosophy, which makes deduction no whit inferior to induction in the tool-chest of the naturalist.

It seems equally clear that the application of Davis's method of thinking about land forms is of great worth in the training of young students. That method is based on the exercise of the imagination, the highest faculty of the mind; it is the faculty of seeing things as they are and not as they appear to be. To develop it in the youth of school and college is the most precious privilege of the teachers, and for this purpose few high school subjects are comparable with the evolutionary treatment of landscapes.

In 1889, five years after his first announcement of the cycle idea, Davis published the most remarkable of all of its many applications. The subject of this study is entitled "The Rivers and Valleys of Pennsylvania." In this masterpiece of acute reasoning and close observation in a complicated terrane he traced the influence of a whole set of differing geological structures on the development of highly varied land forms and of the associated river system. The results of this path-breaking research make this early paper a classic, the conclusions of which stand fast after more than half a century has added to our knowledge of the Pennsylvanian region.

Other broad units of the earth's relief were similarly treated in scores of later papers. At first their author went into the more easily accessible fields which were already covered by reasonably accurate topographic maps: for example, northern New Jersey, southern New England, and Virginia. Then, as a result of many visits to Europe and travels in central Asia, South Africa, Australia, and New Zealand, he tested, far and wide, his art of describing land forms genetically, in terms of structure, process and stage. As he himself expected, he found new complications, but none that could not be fitted into the general scheme, so long as each individual region is affected by the normal climate. Arid regions, however, demanded different treatment, and, aided by the writings of Passarge and others, Davis worked out a scheme for "the desert cycle." His personal inspection of the great topographic changes wrought by mountain glaciation in central France, the Alps, Norway, and our western Cordillera led him to an incomplete but illuminating version of a "glacial cycle," this to include evolutionary stages quite different from, though in some instances analogous to, the systematic stages demonstrated in regions exposed to normal climatic conditions. From his field studies of the New England and other coastlines, supplemented by examination of large-scale maps of continental and island shores the world over, Davis aided by his pupil, F. P. Gulliver, showed how shoreline forms can be systematized and scientifically described in the terms of the "cycle of marine erosion."

Two masterly, advanced courses in physical geography, one on the United States and the other on Europe, claimed the un-fading admiration of those who listened. Illustrated with a host of large-scale topographic maps of States and European countries, these lectures showed the solid worth of Davis's philosophy, though in scholarly fashion he gave full weight to the opinions and methods of other investigators on the two continents. Probably because of the difficulty of adequately reproducing the maps around which the discussion centered, the material of these unique lectures was never published. To spread his gospel Davis relied chiefly on what he used to call "the rapid-fire gun," propagandizing with hundreds of papers, a number of which were written in French and German and printed in Europe. To the teachers in secondary schools he offered his elementary "Physical Geography" (1898) and a second book, "Practical Exercises in Physical Geography" (1908), but the only comprehensive statement of his matured philosophy was published in German with the title "Die Erklärende Beschreibung der Landformen" (1912). In English we have a convenient assembly of twenty-six among the more important papers dealing with methods of teaching geomorphology and with the general idea of the erosion cycle. This volume of nearly 800 pages was edited by the late Douglas W. Johnson, fellow member of the National Academy of Sciences, with the title "Geographical Essays" (1909).

Not the least merit of Davis's papers and books is their profuse illustration with block diagrams, which tell his story with extraordinary clarity and conciseness. His sureness of pen-stroke and his sense of values in selecting the essential features of the thousand landscapes he pictured entitle him to the name artist. In this art no geographer nor geologist has ever rivaled him. Everyone who saw him do it marvelled at his simultaneous use of both hands when drawing block diagrams on the blackboard—with amazing speed and practically without erasures.

In 1912 Davis resigned from the professorship of geology which he had held for thirteen years, after having been Harvard's leading geographer for fourteen years. Thus for nearly thirty years he had been a bridge-builder between the two sciences. It was natural that he should be attracted to the problem of coral reefs, which is obviously in the border field. In 1914 a grant from the Shaler Memorial Fund of his university enabled him to visit many islands in the Fiji, New Hebrides, Cook, Loyalty, and Society groups as well as Oahu, New Caledonia, and a long stretch of the Queensland coast inside the Great Barrier Reef of Australia. In 1923 he added to his field experience by travel among the reef-bearing islands of the Lesser Antilles. For twelve years his time was largely spent on the study of his own observations, of the multitude of island charts issued by the hydrographic offices of the world, and on the voluminous literature on the controversial subject of reef origin. At intervals he published the results of his correlations, producing twenty-eight papers and a book on the Antilles. In 1928 there appeared his weighty monograph, entitled "The Coral-Reef Problem," giving his complete views concerning the relative merits of the many hypotheses which have been offered as solutions to the reef problem.

Davis was fascinated by the beauty and apparent cogency of the Darwin-Dana view that atolls and barrier reefs are best regarded as the products of slow subsidence of the foundations on which these structures are built, and at first (1915) thought the subsidence hypothesis to be alone competent in explanation. Later he accepted the idea of "Glacial controls" as useful in accounting for the "platform foundations" and crowning reefs in the marginal areas of the earth's coral-reef zone. His treatment of the problem was dominated by the double principle of deduction and verification, but in the opinion of the present writer Davis failed to give adequate consideration to some of his premises, including the geological dates when the reef foundations were prepared and when the wave-resisting species of corals became abundant in the tropical ocean. Nor was sufficient attention paid to the relatively enormous areas and remarkable flatness of the lagoons inside atoll and barrier reefs—features which are almost universal and not to be expected on the Darwin-Dana hypothesis. It may further be remarked that this hypothesis is not supported by the findings at test bore-holes in Bermuda and at Michaelmas Cay and Heron Island inside the Australian Great Barrier Reef.

Notwithstanding such failure to secure the premises on which the author of "The Coral-Reef Problem" based his own conclusions, this book will long remain the Bible for geologists and geographers who need a richly illustrated handbook summarizing the facts known about these marvelous structures of the coral seas, or are interested in the relation of the reef controversy to the fundamental question as to the strength and stability of the earth's crust.

Personal Characteristics

Davis had a wonderful capacity for continuous labor. Great physical endurance helps to explain his keen zest for life as well as his success in systematizing a world-embracing science. It took zeal and courage to attempt wholesale reform of the geography taught before his time; both qualities were confirmed as he saw his heresies become gradually accepted principles. His favorite tool was logic. Although at heart he was capable of deep emotion, he would rarely allow emotion to appear in his writings or in his college lectures. Partly for this reason the writings did not appeal to the general public, nor the lectures to the rank and file of Harvard students. Davis was sometimes severely critical of student or colleague who, in order to lighten style of presentation, used simile, metaphor, or other figure of speech which could in the least obscure orderly expression of the thought. Rigorous with himself, he was rigorous with his students. He detested sloppiness and made disciplined thought and precision the outstanding aims of his courses in both college and graduate school. Yet he was sympathetic with honest endeavor and spent much time and energy helping special students who through no fault of their own, had not been properly prepared for imaginative and logical attack on scientific problems.

By his Quaker upbringing Davis was endowed with a high ethical standard. As we have already noted, his family was forced to leave the Society of Friends, but Davis kept one concrete relic of that early association. Even into old age he addressed each member of his own family with the pronoun "thee." Perhaps this habit of speech was rooted so deeply because of a scene witnessed during his plastic childhood. Then he heard a Quaker boy, fighting with another boy who was not of the Friends, intersperse his blows with the taunt: "Thee little You, thee!" The influence of his forbears was particularly shown in Davis's craving for fairness and justice in the world and in his religious tolerance. He used to say: "Who am I to 'tolerate' anybody's belief? I want to respect it even if I cannot agree." He affiliated himself with the Unitarian church. Two months after his death his last paper, "The Faith of Reverent Science," was published. He there declared his ideal for the human race—progress ever upward "to a truly Christian standard."

KEY TO ABBREVIATIONS USED IN THE BIBLIOGRAPHY

Am. Ac. Pr. = American Academy of Arts and Sciences Proceedings
Am. Assn. Pr. = American Association Proceedings
Am. Assn. Adv. Sc. Pr. = American Association for the Advancement of Science Proceedings
Am. G. = American Geologist
Am. Geog. Soc. Bull. = American Geographical Society Bulletin
Am. Geog. Sp. Pub. = American Geographical Special Publication
Am. Geophys. Tr. = American Geophysical Union Transactions
Am. J. Sci. = American Journal of Science
Am. Met. J. = American Meteorological Society Journal
Am. Nat. = American Naturalist
Am. Ph. Soc. Pr. = American Philosophical Society Proceedings
An. Rep. Astron. Obs. Harvard Coll. = Annual Report of the Director of the Harvard Astronomical Observatory
An. Géog. = Annales de Géographie
Assn. Am. Geog. An. = Association of American Geographers Annals
Atl. Mo. = Atlantic Monthly
Biog. Mem. Nat. Acad. Sci. = Biographical Memoirs, National Academy of Sciences
Boll. R. Soc. Geog. = Bollettino Royal Società Geografica
Boston Soc. N. H. Pr. = Boston Society of Natural History Proceedings
Brit. Assn. Adv. Sci. Rep. = British Association for the Advancement of Science Report
Bull. Volcanologique = Bulletin Volcanologique
Calif. J. Mines and Geol. = California Journal of Mines and Geology
Conn. Sch. Doc. = Connecticut School Document
Ed. Rev. = Educational Review
Eng. Mo. J. = Engineers Monthly Journal
Franklin Inst. J. = Franklin Institute Journal
Geog. Anzeiger = Geographischer Anzeiger
Geog. J. = Geographical Journal
Geog. Rev. = Geographical Review
Geog. Soc. Phila. Bull. = Geographical Society of Philadelphia Bulletin
Geog. Teacher = Geography Teacher
G. Assn. Pr. = Geologists Association Proceedings
G. Mag. = Geological Magazine
G. Rundschau = Geologische Rundschau
G. Soc. Am. Bull. = Geological Society of America Bulletin
G. Soc. Am. Pr. = Geological Society of America Proceedings
Ges. Deutsch. Naturf. u. Ärtze = Gesellschaft Deutscher Naturforscher und Ärtze
Ges. Erdk. Berlin Zs. = Gesellschaft Erdkunde Berlin Zeitschrift
Goldthwaite's Geog. Mag. = Goldthwaite's Geographical Magazine

Harvard Coll. Mus. C. Z. Bull. = Harvard College, Museum of Comparative Zoology Bulletin
Harvard Grad. Mag. = Harvard Graduates' Magazine
Int. Cong. Geol. C. R. = International Congrès Géologique Compte Rendu
Internat. Wochenschr. = Internationale Wochenschrift für Wissenschaft Kunst und Technik
J. Franklin Inst. = Journal of the Franklin Institute
J. G. = Journal of Geology
J. Geog. = Journal of Geography
J. N. E. Waterworks Assn. = Journal of New England Waterworks Association
J. Sch. Geog. = Journal of School Geography
Johns Hopkins Univ. Cir. = Johns Hopkins University Circular
Liverpool G. Soc. Pr. = Liverpool Geological Society Proceedings
Mass. St. Bd. Educ. Ann. Rep. = Massachusetts State Board of Education Annual Report
Meriden Sci. Assn. Tr. = Meriden Scientific Association Transactions
Meteorologische Zeit. = Meteorologische Zeitschrift
Mo. Wea. Rev. = Monthly Weather Review
Nat. Acad. Sci. Pr. = National Academy of Sciences Proceedings
Nat. Geog. Mag. = National Geographic Magazine
Nat. Geog. Mon. = National Geographic Monograph
Nat. Herbart Soc. = National Herbart Society
Nat. Hist. = Natural History
Nat. Research Council = National Research Council
Pan-Am. Geol. = Pan-American Geologist
Pan-Pac. Sci. Cong. Pr. = Pan-Pacific Scientific Congress Proceedings
Pop. Sci. Mo. = Popular Science Monthly
Proc. New Eng. Met. Soc. = Proceedings New England Meteorological Society
Quart. J. Geol. Soc. London = Quarterly Journal of the Geological Society of London
Quart. J. Royal Met. Soc. = Quarterly Journal of the Royal Meteorological Society
R. I. Ed. Pub. = Rhode Island Educational Publication
Roy. Geog. Soc. J. = Royal Geographical Society Journal
San Diego Soc. Nat. Hist. Tr. = San Diego Society of Natural History Transactions
Science = Science
Sci. Mo. = Scientific Monthly
Sci. Prog. = Science Progress
Scottish Geog. Mag. = Scottish Geographical Magazine
Tr. Edin. Geol. Soc. = Transactions of the Edinburgh Geological Society
Tr. N. Z. Inst. = Transactions of the New Zealand Institute

U. S. Dept. Ag. Wea. Bur. Bull. = United States Department of Agriculture Weather Bureau Bulletin
U.S.G.S. Ann. Rep. = United States Geological Survey Annual Report
Van Nostrand's Eng. Mag. = Van Nostrand's Engineering Magazine
Ver. Erdk. Leipzig Mitt. = Verein für Erdkunde Leipzig Mitteilungen
Wash. Acad. Sci. J. = Washington Academy of Sciences Journal

BIBLIOGRAPHY OF WILLIAM MORRIS DAVIS

1880

Banded Amygdules of the Brighton Amygdaloid. Boston Soc. N. H. Pr. 20: 426-8.

1881

Illustrations of the Earth's Surface; Glaciers. (With N. S. Shaler.) Science 2: 581-4, 624-30.
Remarks on the Geology of Mt. Desert, Me. Boston Soc. N. H. Pr. 21: 117-8.

1882

Glacial Erosion. Boston Soc. N. H. Pr. 22: 19-58.
On the Classification of Lake Basins. Boston Soc. N. H. Pr. 21: 315-81. (Abst.) Am. Nat. 16: 1028-9.
The Little Mountains East of the Catskills [N.Y.] Appalachia 3: 20-33.
On the Triassic Trap Rocks of Mass., Conn. and N. J. Am. J. Sci. (3) 24: 345-9.

1883

Becraft's Mountain [Columbia Co., N. Y.] Am. J. Sci. (3) 26: 381-9, map.
Deflective Effect of the Earth's Rotation. Van Nostrand's Eng. Mag. 28: 297-8.
Charcoal as Applied to the Deposition of Gold from Copper and Other Impurities. Franklin Inst. J. 85: 274-87.
The Folded Helderberg Limestones East of the Catskills. Harvard Coll. Mus. C. Z. Bull.: 7 (g s l) 311-29, map.
Lake Bonneville. Science 1: 570.
The Nonconformity at Rondout, N. Y. Am. J. Sci. (3) 26: 389-95, map.
Lakes and Valleys in Northeastern Pennsylvania. Science 1: 304-5.
The Origin of Cross Valleys. Science 1: 325-7, 356-7.
On the Relations of the Triassic Traps and Sandstones of the Eastern United States. Harvard Coll. Mus. C. Z. Bull.: 7 (g s l): 249-390.
The Structural Value of the Trap Ridges of the Connecticut Valley. Boston Soc. N. H. Pr. 22: 116-24.

1884

The Distribution and Origin of Drumlins. Am. J. Sci. (3) 28: 407-16.
Drumlins. Science 4: 418-20.
Gorges and Waterfalls. Am. J. Sci. (3) 28: 123-32.
How Do the Winds Blow Within the Storm-Disk? Science 3: 402-3.
Light in the Deep Sea. Science 4: 94.
Local and Tropical Weather Conditions. Am. Met. J. 1: 245-7.
Meteorological Charts of the North Atlantic. Science 3: 654-7.
The Older Wind-Charts of the North Atlantic. Science 3: 593-7.
[On the Definition of a Tornado] Am. Met. J. 1: 159-60.
Paleozoic High Tides. Science 3: 473-4.
Rainfall Maps. Am. Met. J. 1: 302-3.
The Relation of Tornadoes to Cyclones. Am. Met. J. 1: 121-7.
Ueber Samum und Böen. Meteorologische Zeit. 1: 243-5.
Whirlwinds, Cyclones, and Tornadoes. 90 pp., Boston, Lee and Shepard, Publishers.
The Winds and Currents of the Equatorial Atlantic. Am. Met. J. 1: 48-56.

1885

The Deflective Effect of the Earth's Rotation. Am. Met. J. 1: 516-24.
Earthquakes in New England. Appalachia 4: 190-4.
Geographic Classification, Illustrated by a Study of Plains, Plateaus, and Their Derivatives. (Abst.) Am. Assn. Pr. 33: 428-32.
The Meteorological Observatory on Blue Hill. Science 5: 440.
Mountain Meteorology. Appalachia 4: Pt. I, 225-44; Pt. II, 327-51.
The Reddish-Brown Ring Around the Sun. Science 5: 455-6.
Reduction of Barometer Readings to Latitude 45°. Am. Met. J. 1: 516-24.
Relation of Tornadoes to Cyclones. Am. Met. J. 1: 48-56.
Terminology of Atmospheric Vapour. Am. Met. J. 2: 6-7.
Winds and Currents of the Equatorial Atlantic. Am. Met. J. 1: 48-56.

1886

American Contributions to Meteorology. Philadelphia, Nov. 19. Reprinted from J. Franklin Inst. 127: 27 pp.
Bishop's Ring Around the Sun. Pop. Sci. Mo. 28: 466-74.
Bishop's Ring During Solar Eclipses. Science 7: 239-40.
Brief Notices of Papers Read Before the Geological Section of the American Association. Am. J. Sci. (3) 32: 319-24.
'Chinook Winds'. Science 7: 55-6.
Chinook Winds of the North-West. Boston Soc. N. H. Pr. 23: 249-50.
Cyclones, Anticyclones and Pericyclones. Am. Met. J. 3: 117-8.
Derivation of the Term "Trade Wind." Am. Met. J. 3: 111-2.
Earthquakes in New England. Appalachia 4: 190-4.

The Festoon Cloud. Science 7: 57-8.

Foreign Studies of Thunder-Storms. Am. Met. J. 2: 489-99, 3: 40-8, 65-6, 69-79.

(With A. McAdie.) Height and Velocity of Clouds. Ann. Rep. Astron. Obs. Harvard Coll. 40: 10.

Mechanical Origin of the Triassic Monoclinal in the Connecticut Valley. (Abst.) Am. J. Sci. (3) 32: 321 (1886) Am. Assn. Pr. 35: 224-7 (1887) Boston Soc. N. H. Pr. 23: 329-41 (1887).

On the Methods of Study of Thunder-Storms. Am. Ac. Pr. 21: 336-47.

Mountain Meteorology. Appalachia 4: 225-44, 327-50.

Notes on Studies of Thunderstorms in Europe. First Papers: Am. Met. J. 3: Reprinted. 11 pp.

A Recent Ice-Storm. Science 7: 190.

The Recent Cold Wave. Science 7: 70-1.

Relation of the Coal of Montana to the Older Rocks. U. S. 10th Census. 15: 697-712.

(With Shaler, N. S., and Harris, T. W.). A series of twenty-five colored geological models and twenty-five photographs of important geological objects, each accompanied by a letter-press description. [D. C. Heath & Co.]

The Structure of the Triassic Formation of the Connecticut Valley. Am. J. Sci. (3) 32: 342-52.

Temperature Diagrams. Am. Met. J. 2: 169-75.

The Temperature of Mediterranean Seas. Am. Met. J. 3: 49.

A Thunder-Squall in New England. Science 7: 436-7.

Thunderstorms in New England in the Summer of 1885. Am. Ac. Pr. 22: 14-58.

Weather Prediction in New Zealand. Am. Met. J. 3: 103-5.

Winter on Mount Washington. Science 7: 40-2.

1887

Advances in Meteorology. Science 9: 539-41.

The Classification of Lakes. Science 10: 142-3.

The Foehn in the Andes. Am. Met. J. 3: 507. Reprinted.

The Height of Cumulus Clouds. Am. Met. J. 3: 492-4.

Instruction in Geological Investigations. Am. Nat. 21: 810-25.

Land and Sea Breezes. Am. Met. J. 4: 2 pp.

Snow as a Cause of Cold Weather. Am. Met. J. 3: 389-90.

Water-Vapor and Radiation. Am. Met. J. 3: 443-5.

1888

A Classification of the Winds. Am. Met. J. 4: 512-9.

Geographic Methods in Geologic Investigation. Nat. Geog. Mag. 1: 11-26.

Local Weather Predictions. Am. Met. J. 4: 409-12.

[On the Use of Meteorological Maps in Schools.] Am. Met. J. 4: 489-92.
The Structure of the Triassic Formation of the Connecticut Valley.
U.S.G.S. Ann. Rep. 7: 455-90.
Synclinal Mountains and Anticlinal Valleys. Science 12: 320.
The Topographic Map of New Jersey. Science 12: 206-7.
Two Chapters on the Physical Geography and Climate of New England.
Cambridge, 10 pp.
Wasp-Stings. Science 11: 50.

1889

The Ash Bed at Meriden and Its Structural Relation. Meriden Sci.
Assn. Tr. 3: 23-30.
The Contoured Map of Massachusetts. Science 14: 422-3.
The Faults in the Triassic Formation Near Meriden, Conn. Harvard Coll.
Mus. C. Z. Bull. 16 (g s 2): 61-87.
(With Wood, J. Walter, Jr.) The Geographic Development of Northern
New Jersey. Boston Soc. N. H. Pr. 24: 365-423.
The Glacial Origin of Cliffs. Am. G. 3: 14-8.
The Intrusive and Extrusive Triassic Trap Sheets of the Connecticut
Valley. Harvard Coll. Mus. C. Z. Bull. 16 (g s 2): 99-138.
Methods and Models in Geographic Teaching. Am. Nat. 23: 566-83.
(Abst.), Johns Hopkins Univ. Cir. 8: 62.
Report of Investigation on the Sea-Breeze. Am. Met. J. 6: 4-6.
A River Pirate [Deer Run, Pa.] Science 13: 108-9.
The Rivers and Valleys of Pennsylvania. Nat. Geog. Mag. 1: 183-253.
Some American Contributions to Meteorology. J. Franklin Inst. 127:
104-15; 176-91.

1890

Dr. Hann's Studies on Cyclones and Anticyclones. Science 15: 332-3, 1891,
4-5.
[The Features of Tornadoes and Their Distinction from Other Storms]
Am. Met. J. 7: 433-6.
Ferrel's Convectional Theory of Tornadoes. Am. Met. J. 6: 337-49; 418-63.
Investigations of the New England Meteorological Society in the year
1890. An. Rep. Astron. Obs. Harvard Coll. 21 (Pt. 2): 105-213.
The Iroquois Beach. Am. G. 6: 400.
The Lawrence Tornado of July 26, 1890. An. Rep. Astron. Obs. Harvard
Coll. 31 (Pt. 1): 119-137.
The Level of No Strain. Am. G. 5: 190-1.
Oscillations of Lakes (Seiches). Science 15: 117.
An Outline of Meteorology. Johns Hopkins Univ. Cir. 9: 71-2.
The Rivers of Northern New Jersey, with Notes on the Classification of
Rivers in General. Nat. Geog. Mag. 2: 81-110.

Secular Changes in Climate. Am. Met. J. 7: 67-81.

Structure and Origin of Glacial Sand Plains. G. Soc. Am. Bull. 1: 195-202.

Types of New England Weather Observations and Investigations of the New England Meteorological Society. 1888-1889. Investigation of the Seabreeze. An. Rep. Astron. Obs. Harvard Coll. 21 (Pt. 1): 215-65.

Vertical Components of Motion in Cyclones and Anticyclones. Science 15: 388.

1891

The Catskill Delta in the Postglacial Hudson Estuary. Boston Soc. N. H. Pr. 25: 318-35. (Abst.) J. G. 1: 97-8. (1893)

Cumulus Clouds Over Islands. Am. Met. J. 7: 563-4.

European Weather Predictions. Am. Met. J. 8: 53-8.

The Geological Dates of Origin of Certain Topographic Forms on the Atlantic Slope of the United States. G. Soc. Am. Bull. 2: 545-86.

(With Hiram F. Mills and H. H. Clayton). The Lawrence Tornado. Proc. New Eng. Met. Soc., Am. Met. J. 7: 433-43.

The Lost Volcanoes of Connecticut. Pop. Sci. Mo. 40: 221-35.

The Physical Geography of Southern New England. Johns Hopkins Univ. Cir. 10: 78-9.

The Story of a Long Inheritance—Nebula to Tornado. Atl. Mo. 68: 68-78.

Tornadoes; A Story of Long Inheritance. Johns Hopkins Univ. Cir. 10: 78.

The Triassic Sandstone of the Connecticut Valley. Johns Hopkins Univ. Cir. 10: 79.

(With Loper, S. W.) Two Belts of Fossiliferous Black Shale in the Triassic Formation of Connection (with discussion by C. H. Hitchcock and B. K. Emerson). G. Soc. Am. Bull. 2: 415-30.

Was Lake Iroquois an Arm of the Sea? Am. G. 7: 139-40.

1892

The Ancient Shore-Lines of Lake Bonneville. Goldthwaite's Geog. Mag. 3: 105, 1 pl.

The Appalachian Mountains of Pennsylvania. Goldthwaite's Geog. Mag. 3: 343-50.

The Cañon of the Colorado. Goldthwaite's Geog. Mag. 3: 98-102, 1 pl.

The Catskill Delta in the Post-Glacial Hudson Estuary. Boston Soc. N. H. Pr. 25: 318-35.

The Convex Profile of Bad-Land Divides. Science 20: 245.

On the Drainage of the Pennsylvania Appalachians. Boston Soc. N. H. Pr. 25: 418-20.

Ferrel's Contributions to Meteorology. Am. Met. J. 8: 348-59.

The Folds of the Appalachians. Goldthwaite's Geog. Mag. 3: 251-5.

The Loup Rivers in Nebraska. Science 19: 107-8, 220-1.

Meteorology in the Schools. Am. Met. J. 9: 1-21.

Mirage on a Wall. Am. Met. J. 8: 525-6.

Notes on Winter Thunderstorms. Am. Met. J. 9: 164-70.

Observations of the New England Meteorological Society in the Year[s] 1890 [-91]. An. Rep. Astron. Obs. Harvard Coll. 31: (Pt. 1) 1-93.

Outline of Elementary Meteorology. Cambridge, 13 pp.

The Subglacial Origin of Certain Eskers. Boston Soc. N. H. Pr. 25: 477-99. (Abst.) J. G. 1: 95-96 (1893).

The Theories of Artificial and Natural Rainfall. Am. Met. J. 8: 493-502.

1893

Artificial and Natural Rainfall. Boston Commonwealth, Mar. 26.

The Deflective Effect of the Earth's Rotation. Am. Met. J. 10: 195-8.

The General Winds of the Atlantic Ocean. Am. Met. J. 9: 476-88.

Geographical Illustrations; Suggestions for Teaching Physical Geography Based on the Physical Features of Southern New England. 46 pp. Harvard University, Cambridge, Mass.

Geographic Work for State Geological Surveys. G. Soc. Am. Bull. 5: 604-8.

The Improvement of Geographic Teaching. Nat. Geog. Mag. 5: 68-75.

Memorial of James Henry Chapin. G. Soc. Am. Bull. 4: 406-8.

Note on Winter Thunderstorms. Am. Met. J. 9: 164-70.

Observations of the New England Meteorological Society in the Year[s] 1890 [-91]. An. Rep. Astron. Obs. Harvard Coll. 31: (Pt. 2) 161-2.

[On the Difficulties in Weather Forecasting.] Am. Met. J. 9: 550-3.

The Osage River and the Ozark Uplift. Science 22: 276-9.

Proposed Subjects for Correlated Study by the State Weather Services. Am. Met. J. 9: 68-74.

The Redfield and Espy Period—1830-1855. Rep. of the Inter. Met. Congress held at Chicago, Aug. 1893. U. S. Dept. Ag. Wea. Bur. Bull. 11: 305-16.

The Theory of Cyclones. Am. Met. J. 10: 319-21.

Winter Thunderstorms. Am. Met. J. 9: 238-9.

The Winds of the Indian Ocean. Am. Met. J. 9: 333-43, 2 charts.

1894

The Ancient Outlet of Lake Michigan. Pop. Sci. Mo. 46: 217-29.

(With King, C. F. and Collie, G. L.) Conference on Geography, Chicago, Dec. 28-30, 1892. Report on Governmental Maps for Use in Schools Prepared by a Committee of the Conference on Geography Held in Chicago, Ill., December 1, 1892. New York. 65 pp.

(With Griswold, L. S.) Eastern Boundary of the Connecticut Triassic. G. Soc. Am. Bull. 5: 515-30; (Abst.) Am. G. 13: 145-6; Am. J. Sci. (3) 47: 136-7.

Elementary Meteorology. Boston. pp. xii + 355. Maps and Wdcts.

Facetted Pebbles on Cape Cod, Mass. Boston Soc. N. H. Pr. 26: 166-75; (Abst.) Am. G. 13: 146-7.

Festooned Mammiforms and Pocky Clouds. Am. Met. J. 11: 151-3.

Geographical Work for State Geological Surveys. G. Soc. Am. Bull. 5: 604-8; (Abst.) Am. G. 13: 146.

List of Geographical Lantern Slides. Cambridge. pp. 17.

Papers from the Physical and Geographic Laboratory of Harvard University 11. Reprinted from Ann. Rep. School Comm. of the City of Cambridge for 1893.

Meteorology in the Schools. [Hamilton, N. Y.] pp. 11. School Review ii: 529-39.

Note on Croll's Glacial Theory. Tr. Edin. Geol. Soc. 7: 77-80.

Note on Diffusion of Water Vapor and on Atmospheric Absorption of Terrestrial Radiation. Am. Met. J. 11: 147-51.

An Outline of the Geology of Mount Desert. In Flora of Mount Desert Island, Maine: A preliminary catalogue of the plants growing on Mount Desert and the adjacent islands, by Edward L. Rand and John H. Redfield: 43-71. Cambridge [Mass.].

Physical Geography in the University. J. G. 2: 66-100.

A Speculation in Topographic Climatology. Am. Met. J. 10: 333-43.

A Step Towards Improvement in Teaching Geography. Harvard Teachers Assoc. Leaflet No. 11.

1895

The Absorption of Terrestrial Radiation by the Atmosphere. Science 2: 485-7.

Bearing of Physiography on Uniformitarianism. (Abst.) G. Soc. Am. Bull. 7: 8-11. Am. G. 16: 243-4. Science n.s. 2: 280.

The Development of Certain English Rivers [London]. Geog. J. 5: (no. 2) 127-46, Diagrs.

Notes on Geological Excursions. (Abst.) Science n.s. 2: 744.

The New England States. Boston, pp. 31. Supplement to Frye's Complete Geography. (Also 1902 edition.)

The Physical Geography of Southern New England. Nat. Geog. Soc., Nat. Geog. Mon. 1, No. 9: 269-304.

Physiography as an Alternative Subject for Admission to College: Official Report of the 10th Annual Meeting of the New England Association of Colleges and Preparatory Schools, 38-46.

La Seine, La Meuse et La Moselle. Paris. An. Géog. 5° année. No. 19: 25-9. pl. 2.

Theories of Ocean Currents. Science 2: 824.

Winds and Ocean Currents. Science 2: 342-3.

1896

An Elemenary Presentation of the Titles. Science 3: 569-70.

Large Scale Maps as Geographic Illustrations. J. G. 4: 484-513.

The Outline of Cape Cod. Am. Ac. Pr. 31: 303-32. (Abst.) Am. G. 17: 95-6; Science, n.s. 3: 49-50.

The Peneplain of the Scotch Highland. G. Mag. 3: 525-8.

Plains of Marine and Subaerial Denudation. G. Soc. Am. Bull. 7: 377-98. (Abst.) Am. G. 17: 96-7. Science n.s. 3: 501.

The Physical Geography of Southern New England: The Physiography of the United States (Nat. Geog. Soc.): 269-304, N. Y., American Book Co.

[Physiographic Features of the Middle Susquehanna Region, Pa.] Science n.s. 3: 786-7.

The Quarries in the Lava Beds at Meriden, Conn. Am. J. Sci. (4) 1: 1-13, map.

The Seine, the Meuse and the Moselle. Nat. Geog. Mag. 7: 89-202, 228-38, pls. 21-24, 26.

The Soaring of Birds and Currents of Air. Auk. 13: 92-3.

A Speculation in Topographical Climatology. Am. Met. J. 12: 372-81.

The State Map of Connecticut as an Aid to the Study of Geography in Grammar and High Schools. Conn. Sch. Doc. No. 6: 14 pp.

The State Map of New York as an Aid to the Study of Geography in Grammar and High Schools and Academies. Univ. State of N. Y. Exam. Bull. No. 11: 503-26.

The State Map of Rhode Island as an Aid to the Study of Geography in Grammar and High Schools. R. I. Ed. Pub.: 15 pp.

Josiah Dwight Whitney. Harvard Grad. Mag. 5: 206-9.

1897

The Coastal Plain of Maine. Brit. Assn. Adv. Sci. Rep.: 719-20.

Is the Denver Formation Lacustrine or Fluviatile? Science n.s. 6: 619-21.

(With Curtis, G. C.) The Harvard Geographical Models with Notes on the Construction of the Models by C. G. Curtis. Boston Soc. N. H. Pr. 28: 85-110.

The Present Trend in Geography. A Paper Delivered at the 35th Convocation, Senate Chamber, Albany, New York, June 29: 192-201.

Science in the Schools. Ed. Rev. 13: 429-39.

The State Map of Massachusetts as an Aid to the Study of Geography in Grammar and High Schools. Mass. St. Bd. Educ. 60th Ann. Rep.: 18 pp.

Winds and Ocean Currents. Scottish Geog. Mag. 13: 515-23.

1898

Geography as a University Subject. Scottish Geog. Mag. 14: 24-9.

The Grading of Mountain Slopes. (Abst.) Science n.s. 7: 81.

The Equipment of a Geographical Laboratory. J. Sch. Geog. 2, No. 5: 170-81.
(Assisted by Snyder, W. H.) Physical Geography. 428 pp. Boston.
Systematic Geography. 4th Yearbook Nat. Herbart Soc.: 81-91.
The Triassic Formation of Connecticut. U. S. G. S. Ann. Rep. 1896-97 (1898). 18. Pt. 2: 1-192. pl. 1-20.
Waves and Tides. J. Sch. Geog. 2, No. 4: 122-32.

1899

Balze Per Faglia Nei Monti Lepini. Presso La Società Geografica Italiana. Roma. Bollettino Della Società Geografica Italiana, Fasc. XII: 3-17 (Translated by Fr. M. Pasanisi).
The Circulation of the Atmosphere. Quart. J. Royal Met. Soc. 25. No. 110: 160-9.
Die Cirkulation der Atmosphäre. Das Wetter. Jahr. 16, 201-3, 228-32, 253-59.
Continental Deposits of the Rocky Mountain Region. G. Soc. Am. Bull. 11: 596-601.
The Drainage of Cuestas. G. Assn. Pr., London 16: 75-93.
Un Exemple de Plaine Côtière. La Plaine du Maine (États-Unis) An. Géog. 8: 1-5.
The Geographic Cycle. Geog. J., London 14: 481-504.
"Helen-Wind" Beobachtet In Den Cevennen. Meteorologische Zeit. 16: 124-5.
The Peneplain. Am. G. 23: 207-39. An. Géog., Paris, B.: 289-303, 385-404.
The Rational Element in Geography. Nat. Geog. Mag. 10: 466-73.
The System of the Winds. School World 1: 244-7.

1900

The Basin Deposits of the Rocky Mountain Region. (Abst.) Science n.s. 11: 144.
The Conditions of Formation of Conglomerates, and Criteria for Distinguishing Between Lacustrine and Fluviatile Beds. (Abst. with discussion.) Science n.s. 11: 429-30.
Continental Deposits of the Rocky Mountain Region. (Discussion by S. F. Emmons and W. Cross.) G. Soc. Am. Bull. 11: 596-604; (Abst.) Science n.s. 11: 144.
Les Enseignements du Grand Canyon du Colorado. La Géographie. Bulletin de La Société de Géographie. 15 Janvier. 339-51.
Fault Scarp in the Lepini Mountains, Italy. G. Soc. Am. Bull. 11: 207-16, pls. 18, 19.
The Fresh-Water Tertiary Formations of the Rocky Mountain Region. Am. Ac. Pr. 35: 345-73.

The Geographic Cycle. Paper Read at the VII Intl. Geog. Congress of
  Berlin. Verhandl. d. vii Internat. Geog. Kongr. Berlin. 1899 (1900).
  22-31.
Glacial Erosion in France, Switzerland and Norway. Boston Soc. N. H.
  Pr. 29: 273-322.
Glacial Erosion in the Valley of the Ticino. Appalachia ix: 136-56, pls.
  15, 16.
History of the Cincinnati Anticline (discussion). G. Soc. Am. Bull.
  11: 604-5.
Local Illustrations of Distant Lands. I. A Temporary Sahara. J. Sch.
  Geog. 4. no. 5: 171-5.
Notes on the Colorado Canyon District. Am. J. Sci. (4) 10: 251-9.
Note on River Terraces in New England. (Abst.) G. Soc. Am. Bull.
  12: 483-4.
Peneplains of Central France and Brittany. (Abst.) G. Soc. Am. Bull. 12:
  481-3.
Physical Geography in the High Schools. The School Review 8: 388-404,
  449-56.
The Physical Geography of the Lands. Pop. Sci. Mo. 57: 157-70.
Physiographic Terminology with Special Reference to Land Forms. Sci-
  ence 11: 99.
Practical Exercises in Geography. Nat. Geog. Mag. 11: 62-78.

1901

The Causes of Rainfall. J. N. E. Waterworks Assn. 15: 338-50.
An Excursion in Bosnia, Hercegovinia, and Dalmatia. G. Soc. Am. Bull.
  3: 21-50, pl. 1-4.
An Excursion to the Grand Canyon of the Colorado. Harvard Coll. Mus.
  C. Z. Bull. 38 (g s 5): 107-201; (Abst.) G. Soc. Am. Bull. 12: 483; G.
  Mag. (4) 8: 324; Science n.s. 13: 138.
The Geographical Cycle. Int. Cong. Geog. VII, Verh. Pt. 2: 221-31.
Local Illustrations of Distant Lands. J. Sch. Geog. 5: 85-8.
Maps of the Mississippi River. J. Sch. Geog. 5: 379-82.
Note on River Terraces in New England (Abst.). G. Soc. Am. Bull. 12:
  483-5.
Peneplains of Central France and Brittany. [Abst.] G. Soc. Am. Bull.
  12: 481-3, pls. 44-5.
Practical Exercises in Physical Geography. Proceedings Annual Confer-
  ence N. Y. State Science Teachers Association, Albany, 11 pp.
Les enseignements du Grand Canyon du Colorado. La Géographie, 4:
  339-351.

1902

Base Level, Grade and Peneplain. J. G. 10: 77-111.
Elementary Physical Geography. Boston. Ginn and Co. vviii + 401 pp.
  [Also 1926 printing.]

Field Work in Physical Geography. J. Geog. 1: 17-24, 62-9.

The New England States. Boston. Supplement to Frye's Complete Geography. [Also 1895 edition.]

The Progress of Geography in the School. Chicago, pp. 49. (In National Society for Scientific Study of Education. 1st Yearbook, Pt. II: 7-49.)

River Terraces in New England. Harvard Coll. Mus. C. Z. Bull. 38 (g s 5): 281-346.

Systematic Geography. Am. Ph. Soc. Pr. 41: 235-59.

The Terraces of the Westfield River, Mass. Am. J. Sci. (4) 14: 77-94.

1903

The Basin Ranges of Utah and Nevada. (Abst.) J. G. 11: 1201.

Block Mountains of the Basin-Range Province. (Abst.) Science n.s. 17: 301; G. Soc. Am. Bull. 14: 551; Eng. Mo. J. 75: 153.

The Blue Ridge in Southern Virginia and North Carolina. (Abst.) J. G. 11: 121.

The Blue Ridge of North Carolina. (Abst.) Science n.s. 17: 220.

The Development of River Meanders. G. Mag. (4) 10: 145-8.

Effect of Shore Line on Waves. (Abst.) Science n.s. 15: 88; G. Soc. Am. Bull. 13: 528.

An Excursion to the Plateau Provinces of Utah and Arizona. Harvard Coll. Mus. C. Z. Bull. 42 (g s. 6): 1-50.

The Fresh-Water Tertiaries at Green River, Wyo. (Abst.) Science n.s. 17: 220-221; G. Soc. Am. Bull. 14: 544; J. G. 11: 120.

The Mountain Ranges of the Great Basin. Harvard Coll. Mus. C. Z. Bull. 42 (g s. 6): 129-77.

Practical Exercises in Physiography. J. Geog. 2: 516-20.

The Question of Seminars. Harvard Grad. Mag. pp. 8.

A Scheme of Geography. Geog. J. 22: 413-23. London.

The Stream Contest Along the Blue Ridge. Geog. Soc. Phila. Bull. 3: 213-44.

Walls of the Colorado Canyon. (Abst.) Science n.s. 15: 87; G. Soc. Am. Bull. 13: 528.

1904

A Flat-Topped Range in the Tian-Shan. Appalachia 10: 277-84.

Glacial Erosion in the Sawatch Range, Colo. Appalachia 10: 392-404.

Geography in the United States. Am. G. 33: 156-85; Am. Assn. Adv. Sc. Pr. 53.

The Relations of the Earth Sciences in View of Their Progress in the Nineteenth Century. J. G. 12: 669-87.

A Summer in Turkestan. G. Soc. Am. Bull. 36: 217-28.

1905

The Bearing of Physiography Upon Suess' Theories. Am. J. Sci. (4) 19: 265-73; (Abst.) Int. Cong. Geog. VIII, Rp.: 164.

[The Colorado Canyon. (Abst.)] Science n.s. 21 : 860.

Complications of the Geographical Cycle. Int. Congr. Geog. VIII Rp.: 150-63.

A Day in the Cévennes. Appalachia 11: 110-14, pl. 16-17.

The Geographical Cycle in An Arid Climate. J. G. 13: 381-407.

Glaciation of the Sawatch Range, Colo. Harvard Coll. Mus. C. Z. Bull. 49 (g s 8) : 1-11.

Home Geography. J. Geog. 4: 1-5.

Illustration of Tides by Waves. J. Geog. 4: 290-4.

An Inductive Study of the Content of Geography. n.p. pp. 18.

"A Journey Across Turkestan." Explorations in Turkestan. R. Pumpelly. Expedition of 1903. xii + 324 pp.: Washington, D. C.

Levelling Without Base-Leveling. Science n.s. 21 : 825-8.

An Opportunity for the Association of American Geographers. Amer. Geog. Soc. Bull. 37: 87-6.

Tides in The Bay of Fundy. Nat. Geog. Mag. 16; no. 2, 71-6.

The Wasatch, Canyon and House Ranges, Utah. Harvard Coll. Mus. C. Z. Bull 49 (g s 8) : 17-56.

1906

Biographical Memoir of George Perkins Marsh, 1801-82. 1st Copy: Washington, 1906, pp. [10], 1 pl. 2nd Copy: Washington, 1907. pp. [10], 1 pl. Biogr. Mem. Nat. Acad. Sci. 1909. 6: 71-80 and Portr.

The Colorado Canyon and Its Lessons. Liverpool G. Soc. Pr. 10: 98-102.

The Geographical Cycle in an Arid Climate. Geog. J. 27: 70-73.

Incised Meandering Valleys. Geog. Soc. Phila. Bull. 4, no. 4: 1-11 (182-192).

The Mountains of Southernmost Africa. Am. Geog. Soc. Bull. 38: 593-623.

Observations on South Africa. G. Soc. Am. Bull. 17: 377-450, pls. 47-54.

The Physical Factor in General Geography. The Educational Bi-Monthly 1: 112-22.

The Physiography of the Adirondacks. [Formation of scarps.] Science n.s. 23: 630-1.

The Relations of the Earth Sciences in View of Their Progress in the Nineteenth Century. Cong. Arts and Sci. (St. Louis, 1904) 4: 488-503.

The Sculpture of Mountains by Glaciers. Scottish Geog. Mag. 22: 76-89. (Abst.) Brit. Assn. Rep. 75: 393-4.

Professor Nathaniel S. Shaler. Am. J. Sci. (4) 21: 480-1.

1907

(With Johnson, D. W., and Bowman, Isaiah) Current Notes on Land Forms. Science n.s. 25: 70-3, 229-32, 394-6, 508-10, 833-6, 946-9; 26: 90-3, 152-4, 226-8, 353-6, 450-3, 837-9; 27: 31-33.

Hanging Valleys. Science n.s. 25: 835-6.
Hettner's Conception of Geography. J. Geog. 6: 49-53.
The Place of Coastal Plains in Systematic Physiography. J. Geog. 6: 8-15.
The Terraces of the Maryland Coastal Plain. [Review.] Science n.s. 25: 701-7.

1908

Causes of Permo-Carboniferous Glaciation. J. G. 16: 79-82.
Die Methoden der Amerikanischen Geographischen Forschung. Internationale Wochenschrift für Wissenschaft, Kunst, und Technik, Berlin, Nov. 14.
Practical Exercises in Physical Geography. Boston. pp. 12 + 148. Atlas 50 pp.
The Prairies of North America. Internat. Wochenschr. 2: 1011-18, 1045-50.
Der Grosse Canyon des Colorado. Leipzig, pp. 15. In Ges. Deutsch. Naturf. u. Ärzte 1908 (1909). 1: 157-69.

1909

The Colorado Canyon: Some of the Lessons. Geog. J. 33: 535-40; Am. Geog. Soc. Bull. 41: 345-54; (Abst.) Brit. Assn. Adv. Sci. Rep. 78: 948-9.
Geographical Essays, VI, 777 pp. Boston. Ginn & Co. [Edited by Douglas W. Johnson.]
Glacial Erosion in North Wales. Quart. J. Geol. Soc. London. 65: 281-350, pl. 14.
Der Grosse Cañon des Colorado. Himmel und Erde. 22: 22-41.
Der Grosse Cañon des Colorado-Flusses. Ges. Erdk. Berlin Zs. 3: 164-72.
The Physiographic Subdivisions of the Appalachian Mountain System, and Their Effects Upon Settlement and History (Abst.). Brit. Assn. Adv. Sci. Rep. 78: 761-2.
The Rocky Mountains. pp. 16. In Internat. Wochenschr.
The Systematic Description of Land Forms. Geogr. Journ. Roy. Geogr. Soc. 34: 300-26.
The Valleys of the Cotswold Hills. G. Assn. Pr. 21: 150-2.

1910

American Studies on Glacial Erosion Extrait du Compte Rendu du XI: e Congrès Géologique International. 419-27.
Antarctic Geology and Polar Climates. Am. Ph. Soc. Pr. 49: 200-2.
Deutsche und Romanische Flussterminologie. Geog. Anzeiger. 121-3.
Experiments in Geographical Description. Science n.s. 31: 921-46; Am. Geog. Soc. Bull. 42: 401-35; Scottish Geog. Mag. 26: 561-86.
Notes on the Description of Land Forms. Am. Geog. Soc. Bull. 42: 671-5, 840-4.

Practical Exercises in Physical Geography (Abst.). Int. Cong. Geol. IX, 2: 169-70.
The Theory of Isostasy (Abst. and discussion). G. Soc. Am. Bull. 21 : 777.
Die Umgestaltung der Gebirgsformen durch die Gletscher. (Abst.) Ver. Erdk. Leipzig. Mitt. 28-9.

1911

The Colorado Front Range, a Study in Physiographic Presentation. Assn. Am. Geog. An. 1 : 21-84. (Abst.) Science n.s. 33 : 906.
The Disciplinary Value of Geography. Pop. Sci. Mo. 78 : 105-9, 223-40.
Geographical Descriptions in the Folios of the Geologic Atlas of the United States. (Abst.) G. Soc. Am. Bull. 22 : 736.
Geographical Factors in the Development of South Africa. Journal Race Development 2 : 131-46.
Grundzüge der Physiogeographie. Leipzig pp. [12] + 322. (Also edition 1915-17.)
Notes on the Description of Land Forms. Am. Geog. Soc. Bull. 43 : 46-51, 190-4, 598-604, 679-84, 847-53.
Repeating Patterns in the Relief and in the Structure of the Land. [discussion] (Abst.) G. Soc. Am. Bull. 22 : 717.
Short Studies Abroad—The Seven Hills of Rome. J. Geog. 9 : 197-202, 230-3.

1912

American Studies on Glacial Erosion. Int. Cong. Geol. XI, Stockholm, 1910, 419-27.
The Colorado Front Range. Assn. Am. Geog. An. 1 : 21-84, pl. 1-5.
Die Erklärende Beschreibung der Landformen. XVIII; 565 pp. Leipzig. B. G. Teubner.
L'Esprit Explicatif dans la Géographie Moderne. Paris. Annales de Géographie 21 : 3-21.
Guidebook for the Transcontinental Excursion of 1912. Am. Geog. Soc., New York, 144 pp.
Notes on the Description of Land Forms. Am. Geog. Soc. Bull. 44 : 908-13.
Relation of Geography to Geology (Annual address of the president). G. Soc. Am. Bull. 23 : 93-124.

1913

Dana's Confirmation of Darwin's Theory of Coral Reefs. Am. J. Sci. (4) 35 : 173-88; Nature 90 : 632-4; (Abst.) Science n.s. 37 : 724.
A Geographical Pilgrimage from Ireland to Italy. Assn. Am. Geog. An. 2 : 73-100.
The Grand Canyon of the Colorado. J. Geog. 11 : 310-14.
Human Response to Geographic Environment. Philadelphia. 40 pp.
Kelvin on "Light" and "The Tides." Lect. Dr. Eliot's Five-Foot Shelf of Books. Science 3 : 29-32.

Nomenclature of Surface Forms on Faulted Structures. G. Soc. Am. Bull. 24: 187-216.

The Rhine Gorge and the Bosphorus. J. Geog. 11: 209-15.

Speculative Nature of Geology (Abst.). G. Soc. Am. Bull. 24: 686-7.

Submerged Valleys and Barrier Reefs. Nature 91: 423-4.

Valli Conseguenti e Subseguenti. Roma. pp. 6. Boll. R. Soc. Geog. 12: 1429-32.

1914

The Home Study of Coral Reefs. Am. Geog. Soc. Bull. 46: 561-77, 641-54, 721-39.

Meandering Valleys and Underfit Rivers. Assn. Am. Geog. An. 3: 3-28 [1914?].

Sublacustrine Glacial Erosion in Montana. (Abst.) G. Soc. Am. Bull. 25: 86.

Der Valdarno; eine Darstellungstudie. Berlin, pp. 68, 2 pls. In Zeitschr. Gesellsch. f. Erdk.

1915

Biographical Memoir of John Wesley Powell, 1834-1902. Biog. Mem. Nat. Acad. Sci. 8: 11-83, port.

Biographical Memoir of Peter Lesley, 1819-1903. Biog. Mem. Nat. Acad. Sci. 8: 155-240, port.

The Development of the Transcontinental Excursion of 1912. Mem. Vol. Transcontinental Excursion, 1912. Am. Geog. Soc. 3-7.

The Mission Range, Montana. (Abst.) Assn. Am. Geog. An. 4: 135-6.

The Mission Range, Montana. Nat. Acad. Sci. Pr. 1: 626-8.

The Origin of Coral Reefs. Nat. Acad. Sci. Pr. 1: 146-52.

Physiography of Arid Lands. (Discussion.) Brit. Assn. Adv. Sci. Rep. 84: 365-6.

Preliminary Report on a Shaler Memorial Study of Coral Reefs. Science n.s. 41 :455-8.

Problems Associated with the Origin of Coral Reefs Suggested by a Shaler Memorial Study of the Reefs . . . (Abst.) Science n.s. 41: 569.

Sculpture of the Mission Range, Mont. (Abst.) Science n.s. 42: 685.

A Shaler Memorial Study of Coral Reefs. Am. J. Sci. (4) 40: 223-71.

1916

Clift Islands in the Coral Seas. Nat. Acad. Sci. Pr. 2: 284-8.

Coral Reef Problem. (Abst.) G. Soc. Am. Bull. 27: 46.

Extinguished and Resurgent Coral Reefs. Nat. Acad. Sci. Pr. 2: 466-71.

Marcellus Hartley Memorial Medal. Mo. Wea. Rev. 44: 205-7.

The Mission Range, Montana. Geog. Rev. 2: 267-88.

The Origin of Certain Fiji Atolls. Nat. Acad. Sci. Pr. 2: 471-5.

Practical Exercises on Topographic Maps. J. Geog. 15: 33-41.

The Principles of Geographical Description. Assn. Am. Geog. An. 5. 61-105. [1916.]

Problems Associated with the Study of Coral Reefs. Sci. Mo. 2: 313-33, 479-501, 557-72.

Sinking Islands Versus a Rising Ocean in the Coral-Reef Problem. (Abst.) Science n. s. 43: 721.

(With others) Symposium on the Exploration of the Pacific. Nat. Acad. Sci. Pr. 2: 391-437.

1917

Excursions Around Aix-les-Bains. Cambridge. pp. 27. Published for Y. M. C. A. National War Work Council by Appalachian Mountain Club of Boston.

Les Falaises et La Récifs Coralliens de Tahiti. Paris. pp. 41. An. Géog. 27: 241-84.

The Great Barrier Reef of Australia. Am. J. Sci. 44: 339-50.

Grundzüge der Physiogeographie. Pts. 1-2, Ed. 2. Leipzig. B. G. Teubner, 1915-17. (Also edition 1911.)

The Isostatic Subsidence of Volcanic Islands. Nat. Acad. Sci. Pr. 3: 649-54.

The Structure of High-Standing Atolls. Nat. Acad. Sci. Pr. 3: 473-9.

Sublacustrine Glacial Erosion in Montana. Nat. Acad. Sci. Pr. 4: 695-702.

Topographic Maps of the United States. National Highways Association, Division of Physical Geography, Physiographic Bulletin No. 1: 15 pp.

1918

The Cedar Mountain Trap Ridge Near Hartford. Am. J. Sci. (4) 46: 476-7.

Coral Reefs and Submarine Banks. J. G. 26: 198-223, 289-309, 385-411.

Fringing Reefs of the Philippine Islands. Nat. Acad. Sci. Pr. 4: 197-204.

Geological Terms in Geographical Descriptions. Science n. s. 48: 81-4.

Grove Karl Gilbert. Am. J. Sci. (4) 46: 669-81.

A Handbook of Northern France. Cambridge. Harvard University Press. Pp. 11 + 174, ills.

Metalliferous Laterite in New Caledonia. Nat. Acad. Sci. Pr. 4: 275-80.

Praktische Übungen in physische Geographie. Leipsig etc. B. G. Teubnc-. pp. xii + 115 [1] and atlas of 38 pls.

The Reef-Encircled Islands of the Pacific. J. Geog. 17: 1-8, 58-68, 102-7.

Subsidence of Reef-Encircled Islands. G. Soc. Am. Bull. 29: 71-2 (Abst.), 489-574.

1919

Drainage Evolution on the Yünnan-Tibet Frontier. Geog. Rev. 7: 413-5.

Passarge's Principles of Landscape Description. Geog. Rev. 8: 266-73.

Pumpelly's Reminiscences. Science 49: 61-3.

The Significant Features of Reef-Bordered Coasts. Tr. N. Z. Inst. 51: 6-30.
The Young Coast of Annam and Northern Spain. Geog. Rev. 7: 176-80.

1920

African Rift Valleys. Science 52: 456-8.
The Framework of the Earth. Am. J. Sci. (4) 48: 225-41; (Abst.) G. Soc. Am. Bull. 31, No. 1: 110.
The Function of Geography. Geog. Teacher. 10: 286-91.
Geography at Cambridge University, England. pp. 4. J. Geog. 207-10.
The Islands and Coral Reefs of Fiji. Geog. J. 55: 34-45, 200-20, 377-88, 6 pls.
The Penck Festband. (A Review.) Geog. Rev. 10: 249-61.
The Small Islands of Almost-Atolls. Nature 105: 292-3.

1921

The Coral Reefs of Tutuila, Samoa [New York] pp. 7. Science 53: 559-65.
Features of Glacial Origin in Montana and Idaho. Assn. Am. Geog. An. 10: 75-147, 16 figs.
Lower California and Its Natural Resources. (A review.) Geog. Rev. 11, No. 4: 551-62.

1922

The Barrier Reef of Tagula, New Guinea. Assn. Am. Geog. An. 12: 97-151, pl. 24.
Coral Reefs of the Louisiade Archipelago. Nat. Acad. Sci. Pr. 8: 7-13.
Deflections of Streams by Earth Rotation. Science n. s. 55: 478-9.
Dixey's Physiography of Sierra Leone. Philadelphia, pp. 1-11. Geog. Soc. Phila. Bull. 20: 131-41.
Faults, Underdrag and Landslides of the Great Basin Ranges. G. Soc. Am. Bull. 33, No. 1: 92-6, 2 figs.
Geological Overthrusts and Underdrags. (Abst.) Science n. s. 55: 493.
A Graduate School of Geography. Science 56: 121-33.
Memoir of Frederic Putnam Gulliver. Assn. Am. Geog. An. 11: 112-6.
Peneplains and the Geographical Cycle. G. Soc. Am. Bull. 33, No. 3: 587-98.
The Reasonableness of Science. Sci. Mo. 15: 193-214.
Topographical Maps of the United States. Sci. Mo. 15, No. 6: 557-60.

1923

The Cycle of Erosion and the Summit Level of the Alps. J. G. 31: 1-41.
The Depth of Coral-Reef Lagoons. Nat. Acad. Sci. Pr. 9: 296-301.
Drowned Coral Reefs South of Japan. Nat. Acad. Sci. Pr. 9: 58-62.
The Explanatory Description of Land Forms. Alfred Hettner: Die Oberflächenformen des Festlandes. [Review] Geog. Rev. 13: 318-21.

The Halligs, Vanishing Islands of the North Sea. Geog. Rev. 13: 99-106.
The Island of Oahu. J. Geog. 22, No. 9: 354-7, 3 figs.
The Marginal Belts of the Coral Seas. Am. J. Sci. 5th s., 6: 181-95.
New Zealand Land Forms, C. A. Cotton: Geomorphology of New Zealand.
    Pt. I . . . [Review] Geog. Rev. 13: 321-2.
The Shaping of the Earth's Surface. (A review) Geog. Rev. 13: 599-607.
A Working Model of the Tides. Sci. Mo. 16: 561-72.

1924

Classification of Oceanic Islands. (Abst.) Pan-Am. Geol. 42, No. 4: 319.
    G. Soc. Am. Bull. 36, No. 1; 216-7.
Die Erklärende Beschreibung der Landformen. Deutsch Bearbeitet von
    A. Rühl. Aufl. 2. Leipzig, B. G. Teubner. pp. 31, [I], 565, ills.
The Explanatory Description of Land Forms. Belgrade. pp. 50. Recueil
    de travaux à Jovan Cvijić . . . 287-336, 10 figs.
The Formation of the Lesser Antilles. Nat. Acad. Sci. Pr. 10, No. 6:
    205-11.
Gilbert's Theory of Laccoliths. (Abst.) Wash. Acad. Sci. J. 14, no. 15:
    375.
Modification of Darwin's Theory of Coral Reefs by the Glacial-Control
    Theory. (Abst.) Pan-Am. Geol. 42, no. 1: 73-4; Brit. Assn. Adv.
    Sci. Rep. 92 Meeting: 384-5, 1925.
Notes on Coral Reefs. Pan-Pac. Sci. Cong., Australia, 1923 Pr. 2: 1161-3.
The Oceans. Nat. Hist. 24: 554-65.
The Progress of Geography in the United States. Assn. Am. Geog. An.
    14, no. 4: 159-215.
Shaded Topographic Maps. Science n.s. 60: 325-7.
A Tilted-Up, Beveled-Off Atoll. Science n.s. 60: 51-6, 2 figs.; (Abst.) 59:
    544; Pan-Am. Geol. 42, no. 1: 74; Brit. Assn. Adv. Sci. Rep. 92d
    Meeting: 385, 1925.

1925

The Basin Range Problem. Nat. Acad. Sci. Pr. 11, no. 7: 387-92.
[Comment on Dr. C. W. Kochel's "Abstraktionen in der Geologie."] G.
    Rundschau 16: 313-14.
Laccoliths and Sills. (Abst. with discussion.) Wash. Acad. Sci. J. 15, no.
    18: 414-5. Bull. Volcanologique 2e ann. nos. 5-6: 323-4.
A Roxen Lake in Canada [Lake Timiskaming]. Scottish Geog. Mag. 41,
    no. 2: 65-74, 2 figs.
The Stewart Bank in the China Sea. Science 62: 401-3.

1926

Biographical Memoir of Grove Karl Gilbert, 1843-1918. Nat. Acad. Sci.
    Mem. 21, 5th mem.: 303 pp., 18 figs. and pls. (incl. portraits).

Les Côtes et les Récifs Coralliens de la Nouvelle-Calédonie. Paris. pp. 120, Ills. An. Géog. 34, nos. 189, 190, 191, 192.

Elementary Physical Geography. Boston. Ginn and Co. pp. xviii + 401, Ills. [Also another printing 1902.]

The Lesser Antilles. Am. Geog. Soc., Map of Hispanic America, Pub. 2, 207 pp., 66 figs., 16 pls.

Origin of the Lesser Antilles. (Abst.) G. Soc. Am. Bull. 37, no. 1: 220-1.

Subsidence Rate of Reef-Encircled Islands. Nat. Acad. Sci. Pr. 12, no. 2: 99-105, 2 figs.

The Value of Outrageous Geological Hypotheses. Science n.s. 63: 463-8.

1927

Channels, Valleys, and Intermont Plains. Science n.s. 66: 272-4.

A Migrating Anticline on Fiji. Am. J. Sci. (5) 14: 333-51.

The Rifts of Southern California. Am. J. Sci. (5) 13: 57-72, 6 figs.

1928

The Coral-Reef Problem. Am. Geog. Sp. Pub. no. 9, 596 pp., 227 figs. and pls., New York.

Die Entstehung von Korallenriffen: Ges. Erdk. Berlin Zs. nos. 9-10: 359-91, 13 figs.

The Formation of Coral Reefs. Sci. Mo. 27, no. 4: 289-300, 6 figs.

1929

Geological Map of New Mexico. (Comments on Darton's Map.) Science n.s. 70: 68-70.

Wharton's and Darwin's Theories of Coral Reefs. Sci. Prog. 93: 42-56.

1930

The Desert of the Great Southwest. Harvard Grad. Mag. June, 1930.

The Earth As a Globe. J. Geog. 29: 330-44.

[Elementary Physical Geography.] [In Japanese.] pp. [446]. Ills. Various paging. (Also editions in English; 1902, 1926.)

(With Brooks, Baylor.) The Galiuro Mountains, Arizona. Am. J. Sci. (5) 19: 89-115, 9 figs.

(With Daly, Reginald Aldworth.) Geology and Geography, 1858-1929: The Development of Harvard University, S. E. Morison ed., Chap. 19, pp. 307-328, 4 pls, port., Cambridge, Mass., Harvard University Press.

Origin of Limestone Caverns. G. Soc. Am. Bull. 41: 475-628, pls. 7-8.

The Peacock Range, Arizona. G. Soc. Am. Bull. 41: 293-313.

Periodicity in Desert Physiography. (Abst.) Pan-Am. Geol. 53: no. 4, 320.

Physiographic Contrasts, East and West. Sci. Mo. 30: nos. 5 and 6, 395-415, 501-519, 7 figs., 4 pls.

[Practical Exercises in Physical Geography.] [In Japanese.] n.p. pp. [17]. Ils. and atlas of 43 pp. (English edition, 1908.)

Preparation of Scientific Articles. Science 72: 131-4.

Rock floors in Arid and in Humid Climates. J. G. 38: no. 1, 1-27; no. 2, 136-58, 7 figs.

1931

Clear Lake, California. (Abst.) Science n.s. 74: 572-3.

(With Putnam, William Clement, and Richards, George Lambert, Jr.) Elevated Shore Lines of Santa Monica Mountains. (Absts.) Pan-Am. Geol. 54, no. 2: 154; G. Soc. Am. Bull. 42, no. 1: 309-10.

Nature of Geological Proof, Or How Do You Know You Are Right? (Abst.) Pan-Am. Geol. 55, no. 55: 357-8.

(With Killingsworth, Cecil.) Origin of Caverns. (Abst.) G. Soc. Am. Bull. 42, no. 1: 308-9.

The Origin of Limestone Caverns. Science 73: 327-31.

The Peacock Range, Arizona. G. Soc. Am. Bull. 41, no. 2: 293-313, 7 figs.; (Absts.) Pan-Am. Geol. 53, no. 4: 313; 54, no. 2, 152; G. Soc. Am. Bull. 42, no. 1: 308.

Remarks on Arid Pediments. (Abst.) Pan-Am. Geol. 56, no. 3: 236.

The Santa Cataline Mountains, Arizona. Am. J. Sci. (5), 22: 289-317, 6 figs.; (Absts.) Pan-Am. Geol. 55, no. 5: 372-3; G. Soc. Am. Bull. 43, no. 1: 235, 1932.

Shore Lines of the Santa Monica Mountains, California. (Absts.) G. Soc. Am. Bull. 43, no. 1: 227; Pan-Am. Geol. 55, no. 5: 362-3.

Undertow and Rip Tides. Science n.s. 73: 526-7.

1932

Basin Range Types. Science n.s. 76: 241-5.

Glacial Epochs of the Santa Monica Mountains, California. Nat. Acad. Sci. Pr. 18, no. 11: 659-665, 8 figs.

Piedmont Bench Lands and Primärrümpfe. G. Soc. Am. Bull. 43, no. 2: 399-440, 10 figs.; (Absts.) Pan-Am. Geol. 58, no. 1: 68; G. Soc. Am. Bull. 44, pt. 1: 154.

A Retrospect of Geography. Assn. Am. Geog. An. 22: 211-30.

1933

Geomorphogeny of the Desert. (Abst.) Pan-Am. Geol. 5: 374-5.

Glacial Epochs of the Santa Monica Mountains, California. (Absts.) G. Soc. Am. Bull. 44, no. 5: 1041-1133, 26 figs., 16 pls.; G. Soc. Am. Pr. 304-5; Pan-Am. Geol. 59, no. 4: 306-7.

Granite Domes of the Mojave Desert, California. San Diego Soc. Nat. Hist. Tr., 7, no. 20: 211-58, 34 figs., 4 pls.

The Lakes of California. Calif. J. Mines and Geol. 29, nos. 1, 2: 175-236, 29 figs. 1 pl., map.

Submarine Mock Valleys. Am. Geophys. Tr., 14th Ann. Mtg.: 231-4; Nat. Research Council, June; (Absts.) Pan-Am. Geol. 59; no. 4: 307-8.

Work of Sheetfloods. (Abst.) G. Soc. Am. Bull. 44, pt. 1 : 83.

1934

The Faith of Reverent Science. Sci. Mo. 38: 395-421.

Gardiner on "Coral Reefs and Atolls." (A discussion.) J. G. 42, no. 2: 200-17.

The Long Beach Earthquake. Geog. Rev. 24, no. 1: 1-11, 6 figs.

Submarine Mock Valleys. Geog. Rev. 24, no. 2: 297-308; G. Soc. Am. Pr. 306.

1935

(With Maxson, John Haviland.) Valleys of the Panamint Mountains, California. (Abst.) G. Soc. Am. Pr. 1934: 339.

1936

Geomorphology of Mountainous Deserts. 16th In. Geol. Cong. 1933, Rept. vol. 2: 703-14.

1938

Sheetfloods and Streamfloods. G. Soc. Am. Bull. 49, no. 9: 1337-1416, 15 pls., 33 figs.